SpringerBriefs in Energy

SpringerBriefs in Energy presents concise summaries of cutting-edge research and practical applications in all aspects of Energy. Featuring compact volumes of 50 to 125 pages, the series covers a range of content from professional to academic. Typical topics might include:

- A snapshot of a hot or emerging topic
- A contextual literature review
- A timely report of state-of-the art analytical techniques
- An in-depth case study
- A presentation of core concepts that students must understand in order to make independent contributions.

Briefs allow authors to present their ideas and readers to absorb them with minimal time investment.

Briefs will be published as part of Springer's eBook collection, with millions of users worldwide. In addition, Briefs will be available for individual print and electronic purchase. Briefs are characterized by fast, global electronic dissemination, standard publishing contracts, easy-to-use manuscript preparation and formatting guidelines, and expedited production schedules. We aim for publication 8–12 weeks after acceptance.

Both solicited and unsolicited manuscripts are considered for publication in this series. Briefs can also arise from the scale up of a planned chapter. Instead of simply contributing to an edited volume, the author gets an authored book with the space necessary to provide more data, fundamentals and background on the subject, methodology, future outlook, etc.

SpringerBriefs in Energy contains a distinct subseries focusing on Energy Analysis and edited by Charles Hall, State University of New York. Books for this subseries will emphasize quantitative accounting of energy use and availability, including the potential and limitations of new technologies in terms of energy returned on energy invested.

More information about this series at http://www.springer.com/series/8903

Michel Noussan · Manfred Hafner ·
Simone Tagliapietra

The Future of Transport Between Digitalization and Decarbonization

Trends, Strategies and Effects on Energy Consumption

Michel Noussan
Future Energy Program
Fondazione Eni Enrico Mattei
Milan, Italy

Manfred Hafner
Future Energy Program
Fondazione Eni Enrico Mattei
Milan, Italy

Simone Tagliapietra
Future Energy Program
Fondazione Eni Enrico Mattei
Milan, Italy

ISSN 2191-5520   ISSN 2191-5539   (electronic)
SpringerBriefs in Energy
ISBN 978-3-030-37965-0   ISBN 978-3-030-37966-7   (eBook)
https://doi.org/10.1007/978-3-030-37966-7

This Springer imprint is published by the registered company Springer Nature Switzerland AG
The registered company address is: Gewerbestrasse 11, 6330 Cham, Switzerland

Foreword

We are at the dawn of a new transport era, characterized by accelerated innovation—particularly digital—and a growing pressure to reduce our CO_2 emissions from all sectors of the economy, including transport. This new era is alive with exciting opportunities, as well as challenges.

Swift policy action is required all across the globe to define and revise as appropriate legislative frameworks to put the sector on a sustainability path. Such frameworks must comprise policy measures to increase the efficiency of transport systems and the share of more sustainable modes to boost the uptake of clean vehicles, vessels, and aircraft, and replace fossil fuels with sustainable alternatives. Countries have also the duty to incentivize the right consumer choices and increase investment.

Digitalization will drive the transition toward a more efficient transport system. It is important that all countries step up efforts to create a functional mobility data space. Policies must support the development of innovative, fully interoperable solutions, and incentivize their uptake. For example, countries need to deploy smart traffic management systems enabled by digitalization in order to increase and better manage the capacity of railways and inland waterways, to make best use of airspace, to reduce congestion, to introduce shared and automated mobility solutions, and to facilitate multimodal transport.

On its side, the European Commission is tackling these challenges head-on, guided by the European Green Deal, which sets the target of making Europe's economy climate-neutral by 2050. Transport is, rightfully, at the core of the European Green Deal and must reduce its emissions by 90% by 2050. Our strategy for sustainable and smart mobility addresses all of these areas and will guide the EU as we seek to reduce the environmental footprint of transport in the coming years and guide it toward climate neutrality by 2050.

This unique volume offers a comprehensive analysis of all these decarbonization and digitalization transformations that are faced by the transport sector globally, shedding light on the key factors shaping the transition toward a global sustainable transport system.

Brussels, Belgium   Adina Vălean
                                                            European Commissioner for Transport

Preface

Transport includes a wide range of subsectors, which are characterized by different features and trends. Passenger and freight mobility demand is fulfilled with different transport modes and solutions depending on the purpose of the trip, the distance, the location, the cost, and other aspects related to specific constraints. The global share of urban population is continuously increasing, and the rising mobility demand in cities is calling to effective solutions to limit the impacts of local pollution, congestion, and land use. In parallel, long-haul transport demand is reaching historical highs for both passengers and goods, driving increasing energy consumption. Transport is currently heavily relying on oil, which represents more than 90% of the final energy consumption of the sector. Low-carbon alternatives are available, but given the complexity of the sector, these solutions need to be supported by well-planned policies and strategies that are able to tackle the externalities caused by different transport modes and fuels.

The aim of this book is to provide a complete overview of the current situation of the transport sector, focusing on the ongoing decarbonization and digitalization trends, which are shaping the transition toward a sustainable transport system. An additional focus is given on transport policies, which will be a necessary tool in supporting efficient and effective transport solutions.

Chapter 1 of this book provides an overview on the main aspects related to transport, highlighting the complexity of the sector and the high variability of mobility demand and supply with respect to multiple dimensions, including geography, demography, sectors, technologies, and transport modes. Both passenger and freight transport are at the basis of an effective development of countries and societies, and the sustainability of transport is becoming more and more necessary, due to the rising concerns related to climate change, local pollution, congestions, and safety, especially in large cities all over the world. However, significant differences exist among world regions, since cultural, economic, historical, political, and geographical aspects are crucial in the development of transport modes and infrastructures.

Chapter 2 is focused on the available solutions to decarbonize the transport sector, considering both the powertrain technologies and the issues related to the energy supply chains. Electricity, hydrogen, and biofuels are the main alternative sources for transport systems, and they will be analyzed and compared by considering the state of the art of the technologies of each pathway and the potential future development. Passenger and freight transport have specific features, and so does each different transport mode, resulting in the need to evaluate dedicated applications based on the technical and economic conditions of each technological solution. Moreover, variable conditions across world regions may impact the sustainability of each pathway, particularly in relation to the current and expected power generation mix that varies from a country to another. Opportunities and challenges will be discussed to provide to the readers a clear vision on the strengths and weaknesses of each solution.

In parallel with the policy-driven trend of decarbonization, there is a bottom-up trend related to digital technologies, which are the subject of Chap. 3. Potential game changers include Mobility-as-a-Service, shared mobility, autonomous vehicles, and other effects of extra-sector digital technologies (e.g., online platforms, virtualization, e-commerce). At the same time, digitalization interacts with other mobility trends, and the future effect on the demand for transport and the modal share of travelers is far from being clear. The adoption of various solutions will have a large variability from a country to another, and their success will be based both on their capacity to deliver convenient and reliable services to customers at a lower cost, and on their sustainability in terms of environmental impact at local and global scale, urban traffic, and safety aspects. Policies and regulations will have a crucial role in fostering the deployment of effective solutions based on available technologies.

For this reason, the last chapter of this book Chap. 4 is devoted to present different transport policies worldwide, to describe the different solutions that are available to support the transition toward a sustainable transport system. A specific focus has been given to decarbonization solutions, while at the same time discussing also other critical aspects in transport planning, including local pollution, congestions, and land use. Individual countries will have particular and specific advantages and disadvantages with regards to the implementation of each individual measure. There is therefore no perfect policy combination for ensuring effective decarbonization of the transport sector, but the different policies are focused on the key areas in which countries have tried, are trying, and must continue to try to tackle the ongoing climatic change.

Presenting a comprehensive overview of the issue, this book aims to be accessible to a wide readership of both academics and professionals working in different domains dealing with energy and transport, as well as to general readers interested in the subject.

Support from the Fondazione Eni Enrico Mattei (FEEM) in realizing this book and financing its open access is gratefully acknowledged. Founded in 1989, FEEM is a nonprofit, policy-oriented, international research center, and a think tank producing high-quality, innovative, interdisciplinary, and scientifically sound research on sustainable development. It contributes to the quality of decision-making in public and private spheres through analytical studies, policy advice, scientific dissemination, and high-level education. Thanks to its international network, FEEM integrates its research and dissemination activities with those of the best academic institutions and think tanks around the world.

Within FEEM, the Future Energy research Program (FEP), where this book has been conceived and elaborated, aims to carry out interdisciplinary, scientifically sound, prospective, and policy-oriented applied research, targeted at political and business decision-makers. This aim is achieved through an integrated quantitative and qualitative analysis of energy scenarios and policies. This interdisciplinary approach puts together the major factors driving the change in global energy dynamics (i.e., technological, economic, financial, geopolitical, institutional, and sociological aspects). FEP applies this methodology to a wide range of issues such as energy demand and supply, infrastructures, market analyses, socioeconomic impacts of energy policies.

The authors are thankful to Paolo Carnevale and Luca Farinola, respectively, Executive Director and Financial Director of FEEM, for their strong support in the realization of this book.

Milan, Italy   Michel Noussan
March 2020   Manfred Hafner
                                                                                        Simone Tagliapietra

Contents

About the Authors

Michel Noussan is senior research fellow at Fondazione Eni Enrico Mattei (FEEM) Future Energy Research Program and affiliate professor of Sustainable Transport at SciencesPo Paris School of International Affairs (PSIA). He has been researcher and university lecturer in the domain of energy systems analysis, and he has a track record of several publications in international journals and conferences. He holds a Ph.D. in energy engineering at Politecnico di Torino.

Manfred Hafner is professor of International Energy Economics and Geopolitics teaching, among others, at The Johns Hopkins University School of Advanced International Studies (SAIS Europe) and at SciencesPo Paris School of International Affairs (PSIA). He is also the coordinator of the Future Energy research Program of the Fondazione Eni Enrico Mattei (FEEM). During his over 30-year working experience, he has extensively consulted for governments, international organizations, and the energy industry all over the world.

Simone Tagliapietra is senior research fellow at the Fondazione Eni Enrico Mattei (FEEM) Future Energy research Program and Research Fellow at the Università Cattolica del Sacro Cuore and at Bruegel. He is also adjunct professor of Global Energy and Environment Fundamentals at The Johns Hopkins University School of Advanced International Studies (SAIS Europe). He is the author of Global Energy Fundamentals (Cambridge University Press, 2020).

Chapter 1
The Evolution of Transport Across World Regions

Abstract This chapter aims at providing an overview of the multiple aspects involved in passenger and freight transport, which are the base for the understanding of the energy consumption of the sector, as well as for the current trends and prospects related to digitalization and decarbonization. A brief historical discussion and some trends will be presented, followed by a description of the main modes and technologies, both for passenger and freight transport, and a final focus on the differences across world regions in mobility patterns and behaviors. The evolution of transport systems has led to very different situations worldwide, depending on different strategies related to economic development, geographical limitations and cultural, political and social aspects. Proper sustainable mobility plans need to be based on the specific characteristics of each location, and the integration between different governance levels is of utmost importance to improve the reliability, affordability, and energy performance on the entire transport system.

1.1 Introduction

Transport has evolved in history, following a wide range of drivers, which changed how, how much, when, and why people moved and transported goods between places. Mobility demand has always been driven by the need to access opportunities, related to work, services, shopping or leisure, depending on the specific historical and cultural context.

1.1.1 A Brief Historical Perspective

The history of transport has seen a significant evolution over the centuries, both on the causes of mobility demand and on the available modes, that in turn had an impact on the distance that people and goods could travel. People travel to access opportunities, and the share of each activity has evolved in time, with significant differences across societies worldwide, as well as between urban and rural contexts.

© The Author(s) 2020   1
M. Noussan et al., *The Future of Transport Between Digitalization
and Decarbonization*, SpringerBriefs in Energy,
https://doi.org/10.1007/978-3-030-37966-7_1

An example of the evolution of passenger transport modes can be seen in Fig. 1.1, which depicts the average daily distance travelled by the US citizen in the last century. It is interesting to notice how the availability of different technologies has led to a significant and continuous increase of the average distance travelled, and at the same time, new technologies have led to a decrease in the need of walking. Moreover, average daily travel times remain more or less constant, around a total average of one hour per day, leading to the increase of the distance and area to which people have access to in their daily activities. As a result, cities become larger and continue to attract more and more people worldwide, leading to an urbanization trend for which transport will become crucial.

One of the main impacts of transport is related to the energy required to satisfy the mobility demand. The transport sector currently accounts for almost 30% of the world final energy consumption (IEA, 2018b), reaching 32,494 TWh (2794 Mtoe) in 2017, with a 43% increase from the 22,771 TWh (1958 Mtoe) of 2000. At that time, oil represented almost 97% of the transport energy mix, and today, it slightly decreased to 92% thanks to an increased penetration of electricity (mainly in rail services), biofuels, and natural gas. Still, the transport sector remains today the less diversified, and therefore, there are increasing efforts to try to enhance the use of different low-carbon alternatives to oil products.

A closer look at the evolution of transport energy consumption in the last decades (see Fig. 1.2) highlights its continuous increase, with almost a threefold growth from 1971 to 2015, higher than industry consumption (around +80% increase) or residential consumption (roughly +90%). The chart shows also the clear increase of the share of diesel, which is slowly reaching gasoline in the share of consumption by fuel.

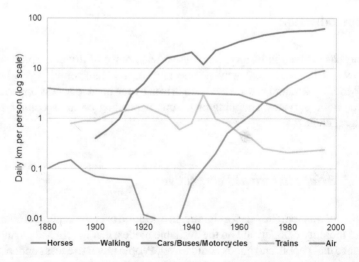

Fig. 1.1   US travel per capita per day by all modes. Authors' elaboration from Ausubel, Marchetti, and Meyer (1998)

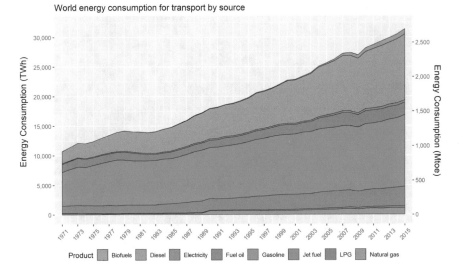

World energy consumption for transport by source

Fig. 1.2 World energy consumption for transport by fuel. Authors' elaboration from IEA (2017)

Transport includes a large variety of subsectors that have peculiar characteristics, as will be described in detail in the following sections. Also, the evolution of transport is tightly related with the urbanization trend worldwide, and mobility planning in cities includes additional aspects related to local pollution, congestions, and safety. It is not trivial to analyze the difference between extra-urban and urban transport, since there are few data specifically related to urban transport at world scale. However, some research has been performed on a limited number of cities, to estimate the transport energy consumption per person in cities related to the population density [see Fig. 1.3, authors' elaboration from WHO (2011)]. Although the data refer to some years ago, the hyperbolic relation among these two quantities appears very clearly. An interesting aspect is the strong clustering of the world regions, which in turn can be correlated to multiple factors including political, economic, cultural, and social behaviors. The cities in the US show generally a low density coupled with the highest per capita energy consumption, which is mostly caused by the diffused use of single-passenger large cars and the low use (and often availability) of public transport. Western European cities lay in the middle, while the bottom right part of the chart is showing mainly high-density cities, whose low per capita energy consumption is a result both of relatively low transport needs due to higher density and low income of the citizens leading to lower access to opportunities. At the same time, in densely populated areas, an excessive use of private vehicles would lead to severe congestions, limiting the speed and flexibility of the private car.

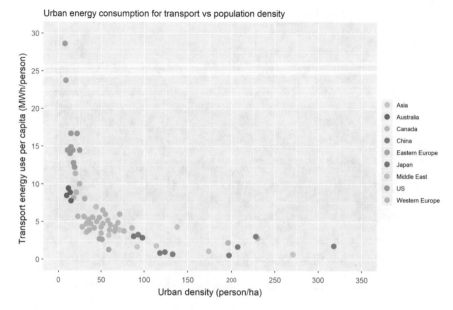

Fig. 1.3 Urban transport energy per capita versus population density. Authors' elaboration from WHO (2011)

1.1.2   Current Situation and Prospects

As discussed above, the evolution of transport demand is related to a number of factors, including the population, the level of GDP, as well as the urbanization. On the other hand, available technologies may unlock additional demand potential, thanks to the availability of mobility solutions at a lower cost for the users or with other advantages, including speed, convenience, and flexibility.

As long as energy consumption is concerned, transport modes can be compared by considering their average specific energy consumption, which can be parameterized on passenger-km (pkm) for passenger transport and on tonnes-km (tkm) for freight transport. Table 1.1 shows some average values for energy consumption of different transport modes, together with their range of variation. These values should be considered with care, since they are affected by a large number of parameters, including type of fuel, vehicle conditions, vehicle load, etc. Moreover, since these figures relate to the final energy consumption of the transport modes, primary energy consumption may differ. For example, transport modes based on electricity show lower specific energy consumption, but the electricity generation may involve additional energy losses in comparison with fossil fuels, depending on the energy source from which it is generated.

Taking in mind these limitations, the values reported in Table 1.1 still provide some interesting evidence: cars and trucks remain among the worst performing transport modes for passenger and freight transport respectively, while the best performing

Table 1.1 Average consumption for transport modes and variation ranges

	Average		Range of variation	
Passenger	kWh/pkm	toe/Mpkm	kWh/pkm	toe/Mpkm
Large cars	0.75	64.7	0.28–1.01	24.3–86.5
Aviation	0.50	43.0	0.29–0.85	24.8–73.1
Cars	0.50	43.0	0.23–0.85	19.8–73.1
Buses and minibuses	0.16	14.1	0.10–0.32	8.3–27.1
2- and 3-wheelers	0.12	10.7	0.10–0.21	8.4–18.5
Rail	0.05	4.1	0.02–0.22	1.5–18.7
Freight	kWh/tkm	toe/Mtkm	kWh/tkm	toe/Mtkm
Medium trucks	0.35	30.5	0.18–0.56	15.3–48.6
Heavy trucks	0.30	25.4	0.22–0.43	18.6–37.2
Rail	0.04	3.5	0.02–0.14	2.1–12.0
Shipping	0.03	2.5	0.02–0.05	2.0–4.0

Authors' elaboration on IEA (2019c)

modes are rail and shipping. Large cars have higher specific energy consumption than aviation, which is generally referred to as the transport mode with the highest environmental impacts. The good performance of rail is partially due to its high electrification, while shipping benefits from the larger volumes, the lower speed, and the lower friction in comparison with land transport.

Besides energy consumption, each transport mode has its own advantages and weaknesses, and the choice of a mode over another may be caused by different reasons. Thus, the transport sector is a complex mix of different modes, and multimodal trips are a common solution for both passenger and freight transport.

The future development of transport needs to be contextualized in an increasingly urbanized world, which will be inhabited by 9.2 billion people by 2040, with 43 megacities of more than 10 million people already in 2030, mainly in Asia and Africa (United Nations, 2018). At the same time, an increasing share of the world population will have access to more services, including private and shared mobility options.

The IEA's World Energy Outlook 2018 presents different future scenarios for the world energy consumption (IEA, 2018b). As far as transport is concerned, its share in final energy consumption by 2040 will still remain similar to the current situation, ranging from 26 to 29%, while the total consumption of the sector will show a higher variation depending on the policies that will be deployed in the future. According to the IEA, while the current policies may lead to an increase of 42% by 2040 compared to the current consumption, with an oil share still locked to 88%, the Sustainable Development Scenario presents a 6% decrease of energy consumption in transport, with oil representing 60% of the 30,703 TWh (2640 Mtoe) estimated for 2040.

The future trends for transport are strongly related to the effectiveness of several policies that may be deployed at different governance levels, as will be better described in Chap. 4. The efforts required for the decarbonization of transport will likely include multiple technological solutions, since no silver bullet appears to be able to tackle the diversified challenges related to the complexity of this sector.

1.2 Passenger Transport

Passenger transport includes a wide variety of activities that range from work commuting, to business trips, tourism, everyday activities, etc. For most trips, there are different potential alternative modes, which may be chosen by the passengers by considering different aspects including price, travel time, comfort, and safety. Passenger transport demand is usually quantified in passenger-km (pkm), which accounts for the transportation of a passenger over a distance of a km. A reliable estimation of passenger transport demand is crucial to perform proper mobility planning strategies at different levels. Figure 1.4 reports the estimations of the International Transport Forum for passenger transport in the world, by highlighting some categories related to modes and distances, as well as the contribution of OECD and non-OECD countries.[1] The chart highlights the very strong increase in transport demand, which is expected to almost triple by 2050, with the strongest contribution coming from

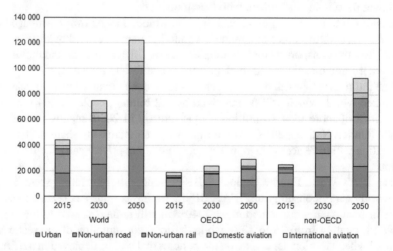

Fig. 1.4 Estimated future demand for passenger transport by type, billion passenger-km. Authors' elaboration on ITF (2019)

---

[1] The Organization for Economic Co-operation and Development (OECD) was established in 1961 as a forum for governments to share experiences and seek solutions to common economic and social problems. As of 2019, the OECD is composed by 36 member countries (https://www.oecd.org/about/members-and-partners/).

developing countries (especially in nonurban road and aviation, i.e., on medium and long distances). The main drivers will be the growing population and the increased well-being, allowing a larger number of people to access different mobility options, especially private cars.

However, attention must be paid to the fact that there is no single categorization for passenger transport, leading to the difficulty of comparing scenarios from different sources due to the different aggregation levels that are used.

This section will present the three main groups that are generally considered (road, rail, and aviation) together with a brief description of the aspects related to active transport (i.e., walking and cycling), which are usually not considered in world statistics but they may be more and more important to develop sustainable alternative to private car in urban contexts. Furthermore, last-mile active solutions may be integrated by micro-mobility electricity-based services, including scooter sharing as well as electric bikes.

1.2.1 Road Transport

Road transport includes all the different motorized vehicles that are available for passengers, including cars, buses, and motorcycles/mopeds. Although with some differences related to the world regions, private cars currently remain the most diffused mode, thanks to their high flexibility and reliability, together with their relatively low cost.

The evolution of private vehicles across countries still shows large inequalities, mostly related to the average income of the citizens. The relation between the vehicle ownership and the gross national income across world countries emerges quite clearly in Fig. 1.5.

It has to be highlighted that the data in the chart are related to the total number of vehicles, including different modes. As a result, for some countries, cars are predominant (in the US, they represent more than 92% of the total registered vehicles), while in other countries, especially in Asia, two- and three-wheelers are the most diffused road transport mode (94% in Vietnam, 73% in India, 54% in Thailand). Similar figures come from research activities based on surveys: Poushter (2015) compared ownership rates of cars, motorcycles, and bikes in 44 countries worldwide, finding large differences from a country to another, as well as depending on the citizens' incomes inside countries.

All road transport vehicles are currently heavily relying on oil products, especially gasoline and diesel. While the latter is generally the most diffused option for heavy-duty vehicles (trucks and buses), for light-duty vehicles the competition is more pronounced, especially in European countries (the US cars almost totally run on gasoline). While diesel engines are generally more efficient, they usually have a higher investment cost and are therefore preferred for users that have a high car usage. Alternative fuels, such as natural gas, LPG and biofuels, are being used in

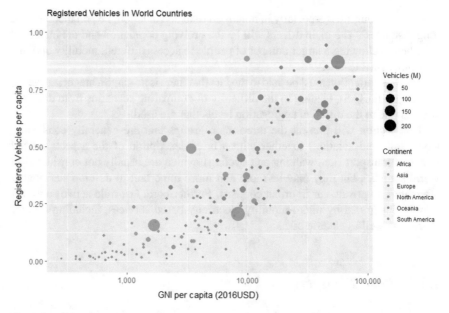

Fig. 1.5 Registered vehicles as a function of gross national income. Authors' calculation on WHO (2018)

some countries, but they are still representing a marginal share of the total energy consumption of the passenger transport.

The energy consumption required for passenger transport on road can show a significant variation based on multiple parameters, including the type of vehicle, the load factor, the average speed, the type of driving, the technology and aging of the vehicle, etc. While average numbers from country statistics are used to estimate the energy consumption of the sector, it has to be noted that the potential variability is significant, especially across countries and in different groups of users inside each country.

There is a significant debate on the road transport total costs considering externalities (Litman & Doherty, 2009), since the vehicle ownership and the operational costs paid by the users are only representing a limited share of the total costs. Other aspects, both internal and external, include the risk of accidents, the cost of congestions (related to the cost of the time that is wasted), energy security, GHG emissions, air pollution, noise, impact of infrastructures, etc. The current taxation system hardly compensates for the external costs of road transport, resulting in an inequitable distribution of costs and benefits. Part of these costs are currently compensated by fuel taxes in some countries, but some alternative solutions are being evaluated, mainly to compensate the decrease of revenues due to the improved efficiency of vehicles and the gradual introduction of electric vehicles. Some US states are considering the possibility of taxing the vehicles based on their usage rather than on fuel consumption, although this would not reflect the issues related to congestion, that would require a pricing both location and time based. The development of digital technologies (onboard sensors, GPS, connectivity, etc.) may support in the future the possibility of dynamic taxation for road transport.

While road transport analyses are generally focused on vehicles, both for economic and energy evaluations, another important aspect that needs to be accounted for is the infrastructure. In fact, road transport needs a proper infrastructure, which in turn requires significant investments and a proper maintenance. The road network has seen a significant evolution in the last centuries, especially in developed countries, in parallel with the development of both passenger and freight transport. The possibility of exploiting a very large and distributed network of roads, from local to highways, makes road transport the only solution that can almost always guarantee a door-to-door service without the integration with other modes. This enhanced flexibility is one of the most important aspects that lead to the success of road transport over other modes (notably railways).

1.2.2 Rail Transport

Rail transport is a significant alternative to road for land transport, thanks to the availability of a separate infrastructure that is not affected by the road network congestions, although it needs a careful planning and management for its optimal operation. Additional advantages of rail over road transport include the higher average speed, especially in urban contexts and in high-speed rail networks, the lower fatality risk for passengers, the better energy efficiency, and lower environmental impacts.

Rail services are generally provided to the users by public or private companies, which allow passengers to travel between specific locations. Competition is possible, but there is the need of a third-party management of the infrastructure to avoid potential collisions or congestions and to optimize the operation and scheduling of the trips. The flexibility of the system is lower than for road transport, although in some countries well-developed rail networks ensure a redundancy that enhances the system flexibility, especially over long distances. However, rail transport is generally part of a multimodal trip that includes other modes for the first- and last-miles.

Rail transport includes different segments that are generally divided into urban railways (i.e., trams and subways), extra-urban conventional railways, and high-speed railways. These segments have different features, targets, and competing transport modes. Urban rail has the potential to support the current urbanization trend by providing an effective alternative to private cars with benefits on local pollution, GHG emissions, congestions, and land use, especially in densely populated districts. High-speed rail, if well planned and operated, can provide an effective substitute for short-haul flights, being aviation one of the most challenging modes to decarbonize.

Globally, around three quarters of conventional passenger rail activity are based on electricity, and the remaining quarter relies on diesel (IEA, 2019c), while high-speed rail and urban rail are totally powered by electricity. According to (IEA, 2019c), the total share of passenger rail transport on electrified tracks is expected to rise to 97% by 2050, with the global activity becoming 2.7 times higher than the current levels.

Globally, rail represented 8% of passenger transport in 2016 (IEA, 2019c), but with an uneven distribution across different world areas. The highest share of passenger transport on extra-urban conventional railways is in Asia, with India accounting for 39% of the total, followed by China with 27% and Japan with 11%. China accounts for about two-thirds of high-speed rail activity, having overtaken both Japan (17%) and the European Union (12%) in the last years. The regional distribution of urban rail activity is more even; China, European Union, and Japan each have around one-fifth of urban passenger rail activity.

While conventional railways have not seen any disruptive improvements in the last century, the evolution of high-speed networks has reached significant penetrations in multiple countries, providing a fast, reliable, and cost-effective alternative for traveling from a city to another. This segment may be further improved in the future by the deployment of alternative technologies that have still a few applications, such as maglev (from "magnetic levitation"), or that are still in a research phase, such as Hyperloop.

Maglev trains are based on a well-known technology that has currently failed in reaching the strong expectations of the past decades, mainly due to the very high capital costs involved. Maglev trains are in commercial operation in six locations in Asia as of 2018 (Maglev.net, 2018), but the only application running at higher speed than normal high-speed trains is the one connecting Shanghai airport with the city center, reaching a top speed of 430 km/h over the 30 km of its length. Another project currently under construction in Japan plans to connect Tokyo and Nagoya, but the benefits provided by halving the travel time come at the cost of a four to five times higher energy consumption in comparison with the current high-speed train connecting those cities (Kingsland, 2018).

Conversely, Hyperloop technology is a new concept that has been proposed for the first time in 2013 by the CEO of Tesla Motors and SpaceX, Elon Musk (SpaceX, 2013). It is based on pressurized capsules that travel in low-pressure tubes at speeds similar or higher than air travel. Different feasibility studies have been developed in recent years, suggesting that this technology could be two to three times more energy efficient per passenger transported than conventional high-speed rail (IEA, 2019c). However, actual figures may vary, and commercial projects are not expected before the mid-2020s (Hyperloop One, 2019).

1.2.3 Air Transport

Aviation is among the most critical transport segments, due to its constantly increasing passenger demand, especially for long-haul flights, and the high energy density that is required. Demand for domestic and international air transport combined is expected to rise from 7 trillion passenger-kilometers in 2015 to 22 trillion in 2050,

Average annual flights per person in selected countries

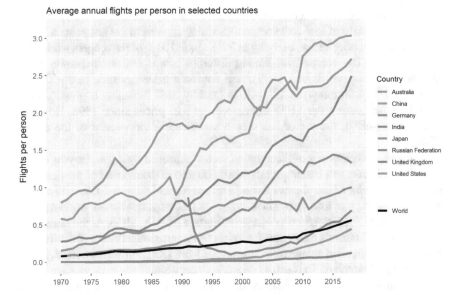

Fig. 1.6 Annual flights per person in selected countries. Authors' calculation on World Bank (2019)

according to (ITF, 2019), especially in developing countries. International air transport passenger demand in China and India alone is expected to increase more than threefold by 2030 and almost sevenfold by 2050.

As for other transport statistics, it is not easy to find reliable and coherent data across world countries with an acceptable historical record. The plot of Fig. 1.6 represents the historical evolution of the average flights per person in some countries, showing the significant differences across developed and developing countries, as well as the growing trend for China and India, which is very similar to the growth at global scale. Unfortunately, there is a caveat: these data are grouped by country based on the nationality of the air carriers rather than on the airports of departure. As a result, while for large countries the values are meaningful, this may not be the case for small countries that are the basis of large international air companies (e.g., the United Arab Emirates, with Etihad and Emirates, or Ireland, with Ryanair). Moreover, while the first decades of air travel were mainly based on national flagship air companies, in the current globalized market air companies may operate flights outside of their home country.

The aviation sector in 2016 accounted for an energy consumption of 2163 TWh (186 Mtoe) for international travels and 1384 TWh (119 Mtoe) for domestic travels (IEA, 2017), all made up of oil products. Aviation accounted for 12% of the total CO_2 emissions of the transport sector in 2016, reaching almost 1 Gt of carbon dioxide (IEA, 2018a). If it was counted as a country, aviation would be ranked sixth for CO_2 emissions from fuel combustion, between Japan (1.1 Gt) and Germany (0.7 Gt).[2]

---

[2]Authors' calculations based on data from IEA (2019b).

The increase of air travel demand has been caused by multiple drivers, including the increasing share of low-cost companies, which showed in the last years a higher growth than the entire sector, carrying in 2018 around 31% of passengers worldwide (ICAO, 2018). The competition among different companies ensures lower fares. According to (Kasper & Lee, 2017), real domestic price per mile in the US has declined by 40% in the years 1990–2016 (and by 36% including bag and change fees), notwithstanding the 110% increase in jet fuel prices since 1998. However, in the same period, the sector has seen a significant improvement of the energy efficiency.

Globally, the efficiency of the aviation sector improved by 2.9% per year during 2000–2016, thanks to a better aircraft utilization and to the renewal of the fleet. The average load factor of the planes reached a record level of 82% in 2018, with a slight increase from the previous year. The average load factor varies across world regions, ranging from 71.8% for Africa to 84.5% for Europe (ICAO, 2018). At the same time, the average fuel burn of new aircraft models fell approximately 45% from 1968 to 2014, or a compounded annual reduction rate of 1.3% (Kharina & Rutherford, 2015), but with significant variations across decades. Energy efficiency measures are driven by the significant cost of fuel in airline operation (reaching a share of roughly 20%), which drives investments in better aircraft designs and lighter materials.

1.2.4   Active Modes

Active transport modes include all the modes that do not require an external energy source for passenger transport, primarily walking and cycling. These transport modes have never been considered in the energy statistics, since they are not related to the energy consumption of any fuel, and at the same time, they represent a marginal share when considering passenger-km. However, they make up a non-negligible share of the number of trips in our everyday life, and they have the potential of replacing a significant number of short motorized trips, especially in densely populated cities. Any proper sustainable mobility planning strategy should be defined by maximizing the contribution of active modes, especially in first- and last-mile solutions. Active transport can also help in preventing the deaths attributable to physical inactivity, which have been estimated to reach 3.2 million worldwide on a yearly basis (WHO, 2011). Moreover, being affordable by virtually everyone, they are the most equitable of all transport modes (Buehler & Pucher, 2012). They also provide advantages related to urban congestion, noise, air pollution, and use of land space.

Due to the lack of a consequential fuel consumption, there is no reliable accounting of the total amount of trips done on foot or by bike on a global level, although some numbers are available on a country basis or for selected cities, calculated with different estimation methods, usually based on surveys. Some researchers have analyzed walking and cycling figures for some countries by considering the most comparable and detailed data, including the US, the United Kingdom, Denmark, France, Germany, and the Netherlands (Buehler & Pucher, 2012). Considering the percentage of total trips in the last decades, a generalized decrease of active modes is noticeable, although with marked differences across countries, which is related to a corresponding rise of travels by private car. The most recent data (2008) show a 11% share of walking in the US, with values in European countries but Denmark above 20%. Cycling shows a larger variability, with slight variations over time in each country, but huge differences from high-cycling countries (the Netherlands, Denmark, and Germany with 25%, 18%, and 10%, respectively, in 2008) to low-cycling countries (France, UK, and USA with 3%, 2%, and 1%, respectively).

A data-driven estimation of physical activity patterns in different countries has been performed by analyzing data from smartphones' accelerometers, to calculate the daily steps performed by the users (Althoff et al., 2017). The researchers found that the average user recorded roughly 5000 steps per day over an average span of 14 h, but with significant differences across countries (from a low 3500 steps in Indonesia to a high of 6900 in Hong Kong). The findings highlighted the role of the built environment in helping citizens improving their daily activity, with beneficial consequences on public health. It must be noted that this research, based on smartphone data, included mostly high- and middle-income countries, while low-income countries may result in even higher steps per day, although not resulting from a choice but rather due to necessity. Data from smartphone may also be affected by smartphone ownership distribution within each country that could be biased based on gender, age, or income.

1.3   Freight Transport

Freight transport has become more and more important in the last decades. This is mainly due to its integration in the manufacturing supply chain driven by the increase of trade at global scale, thanks to better information and communication technologies and favorable regulations (Rodrigue, 2017). More convenient and cheaper freight transport solutions unleashed the possibility of locating production unites in sites with more favorable conditions, exploiting an increasingly complex distribution of final products as well as input materials. With increasing complexity of final products, supply chains are currently involving multiple intermediate products, shipped from different world regions depending on their market conditions. As a result, freight transport is now the backbone of the industrial system at the base of the world economy, and often transport infrastructures (roads, railways, ports, pipelines) are among the key assets for trade and geopolitics.

The main modes involved in freight transport are water (mainly over sea but also in inland waterways), road, and rail, with a minimum contribution of pipelines and air transport. This latter mode is limited to goods for which speed is a crucial requirement, since the cost is significantly higher than for other competing modes. As reported in Fig. 1.7, freight transport demand, usually measured in tonnes-km, is largely dominated by sea transport, and the total demand is expected to increase more than threefold by 2050 (ITF, 2019). Moreover, sea transport will even increase

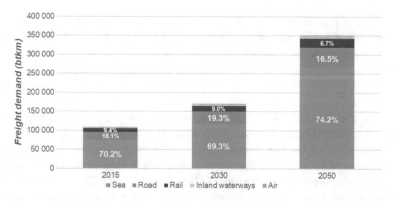

Fig. 1.7 Estimated future demand for world freight transport by type, billion tkm. Authors' elaboration from ITF (2019)

its share, reaching three quarters of the total demand by 2050, with a decrease of both road and rail shares. These projections are heavily dependent on economic growth estimations, and therefore, in the current context of uncertainty, they may be subject to significant variability. Another key factor is the expansion of trade capacity in countries with a substantial potential, notably in Asia.

The energy consumption of freight, as discussed for passenger transport, is largely dominated by oil products, mainly diesel oil for trucks, inland waterways and some rail, and mainly heavy oil for international marine transport. Alternative energy solutions for the decarbonization of the different modes will be discussed in deeper detail in Chap. 2.

In comparison with passenger transport, usually freight transport has less stricter requirements in terms of comfort, speed, and flexibility. Goods can be stored along the trip without significant issues, although in some cases refrigeration is needed to avoid goods deterioration (foods, cold chemicals, etc.). On the other hand, while passengers can face multimodal trips by changing mode by themselves, freight displacement from a mode to another may need longer time and dedicated infrastructure and logistics.

Finally, it is worth noticing that we usually have a more direct knowledge of passenger transport, thanks to our everyday experience, and we are much less familiar with all the aspects involved in freight transport logistics as well as their impacts.

1.3.1 Road Freight

Road freight is the most flexible mode, and it is often used for the last part of the supply chain, although it may cover alone the entire path from production to supply to the final user, especially for short- and medium-distance supply chains. Road freight, thanks to its flexible, convenient, and fast operation, has contributed to the evolution of the goods supply chain from the past, when industries and logistic centers were located near the ports or rivers to allow for maritime transport. The relatively low cost of vehicles and the unrestricted access to road infrastructures make easier for new actors to enter the market, resulting in high competition and low margins in comparison with other modes.

Different truck sizes are available depending on the quantity of goods that are delivered, and the availability of a large and diversified vehicles fleet allows for a better management of goods distribution. There is a limited potential for increasing the size of the trucks, and the corresponding weight and volume of goods that can be carried. National regulations usually limit the maximum weight of trucks due to safety reasons and to the effect of heavy vehicles on the road infrastructure, which results in increased damages and maintenance costs. Moreover, the energy consumption of the engines increases strongly with the weight, resulting in unsustainable economic operation (Rodrigue, 2017).

Currently, road freight transport is totally relying on diesel oil, thanks to the better efficiency of the engines, especially when running at constant speed for long distances. Although it accounts for only 20% of global tkm, road freight consumes more than 70% of energy in freight modes (Teter, 2018). The majority of fuel consumption is due to medium and heavy trucks, and although light commercial vehicles represent a marginal share, they are by far the less efficient freight transport mode. The average energy intensity of trucks in 2017 was 0.35 kWh/tkm (30 toe/Mtkm) for medium trucks (but with significant variations depending on the country) and 0.30 kWh/tkm (25 toe/Mtkm) for heavy trucks (IEA, 2019c). Some alternative fuels are gaining interests thanks to their advantages related to the decarbonization pathways (as it will be better discussed in Chap. 2), including biofuels blends, liquified natural gas, hydrogen, or electricity.

The shift to trucks by many transport companies has led to an increase of road freight transport, with consequences on congestions, especially in urban areas. Urban traffic conditions are often reaching the limits of the available infrastructure, especially due to commuting patterns during day hours, and freight transport in cities is competing with the larger number of vehicles for passengers. As a result, congestions lead to delays and lower frequency of delivery, resulting in the need of increasing the vehicles fleet to ensure the same level of service, with even larger consequences on congestion.

The rise of e-commerce is putting additional pressure on last-mile freight transport, due to the increase of the competition and the speed expected by the customers. The increase of last-mile freight services with door-to-door delivery may be compensated by fewer shopping trips by customers, although the trade-off between these two aspects is very context-specific and it is difficult to draw generalized conclusions. However, e-commerce is often causing additional trips that are required also for very small parcels, as well as the more frequent returns of defective or unwanted goods. Advanced algorithms may increase the effectiveness of door-to-door delivery by an optimization of the organization of trips.

1.3.2 Rail Freight

Rail transport has been at the core of the industrial era, supporting the economic development of countries in the US, Western Europe, and Japan. However, in the last decades, it has faced a strong competition from road transport, whose increasing efficiency and flexibility, together with decreasing costs, were the basis for gaining a considerable market share.

Rail is the land transport mode with the highest capacity, since a single wagon can carry up to 100 t, more than the triple of a truck, and multiple wagons are generally connected in the same train, exploiting economies of scale. There is a wide variety of rail vehicles specialized for different purposes. Open wagons (hopper cars) are used for bulk cargo (e.g., minerals or coal), box cars to carry general and refrigerated goods, and tank cars to carry liquids. The development of intermodal transportation has also supported a new class of flat railcars that can carry containers, in combination with trucks or ships. The trend has thus been toward a specialization of freight wagons for different goods, and a single train can often be composed of various types of wagon, although with higher costs for assembling and organizing goods (Rodrigue, 2017).

Due to the high investment costs, rail companies are often nationalized and operating in conditions of monopoly, or in some cases of oligopoly. There are significant constraints related to the limited time slots available, which lead to the need of a rigid schedule and organization. This aspect is even more critical if the tracks are shared with passenger rail transport, the latter being often prioritized due to its higher requirements in terms of speed. For this reason, in the last decades, freight rail has seen a larger development in regions with dedicated tracks. However, the recent shift toward high-speed trains that are mostly operated on separate tracks is opening additional time slots for freight transport on conventional passenger railways.

A critical issue for freight transport over long distances is related to the standardization of gauge in railway networks: although the standard gauge (1.435 m) is diffused on 60% of the global mileage (mainly in North America and Western Europe), freight transport over long distances involving different gauge systems requires changing vehicles, with consequent higher times and costs. This is one of the obstacles hindering the development of rail transport between Asia and Europe. Other factors limiting the interoperation across multiple countries are related to signaling and electrification standards, which often limit the operation across country borders.

Rail freight transport is currently powered by diesel or electricity, with variable shares across world regions. At a global scale, electricity-powered trains carried 48% of the total tonnes-kilometers (tkm) in 2016 (IEA, 2019c), but this share was over 80% for Japan, Russia, and Europe, while North America and South America were heavily relying on diesel. Since electric trains are more efficient, the final energy consumption of freight transport in 2017 (IEA, 2019c) was 25 Mtoe of diesel (291 TWh) and 130 TWh of electricity (equal to 11 Mtoe), although it is important to notice that the electricity production may involve different amounts of primary energy depending on the region, resulting potentially in higher primary energy consumption. The energy intensity of rail freight transport showed a world average of 0.04 kWh/tkm (3.5 toe/Mtkm) in 2017 (IEA, 2019c), with lower values for China and Russia thanks to the high loading of the trains and electrification rates. Potential alternative energy sources for non-electrified rail lines include natural gas, biofuels, hydrogen, or electric batteries, although none of them has shown so far economic viability. However, decarbonization policies such as carbon taxes may change the equation, and pilot projects are already being evaluated in different countries.

1.3.3 Maritime Freight

Maritime freight represents by far the most common mode of freight transport worldwide considering volume, final energy consumption, and GHG emissions. In the last decades, international shipping has been the backbone of globalization, allowing the development of complex products supply chain based on manufacturing sites located in different world regions, to fully exploit the advantages of local conditions (e.g., resource availability, low wages, national regulations).

One of the most significant game changers has been the development of containerization, which led to faster and more standardized port operations, with a better integration with other transport modes (trains, trucks, and inland navigation). Containerships have now annual sailing times of around 70%, while previously bulk carrier ships used to transport different goods needed longer port operations, as high as 75% of the time (Rodrigue, 2017). Containerization also allowed inter-range services, i.e., a continuous loop involving a sequence of different ports with a flexible frequency based on market conditions, usually including a transoceanic service. The main advantage is the possibility of optimizing the use of the ships by increasing their average load factor, although attention must be paid to avoid the risk of empty trips, particularly in backhauls. Recent trends include intermediate hubs, to avoid the need for large ships to deviate from the main marine shipping routes. Figure 1.8 reports

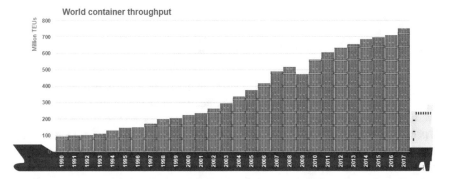

Fig. 1.8 World container throughput, million TEUs. Authors' elaboration from World Bank (2019)

the evolution of world container throughput (i.e., each charge/discharge of containers) measured in twenty-foot equivalent units (TEUs), which are used considering as reference the size of a standard container.

However, maritime trade is still dominated by bulk carriers, which represented around 60% of total tkm in 2015, including ores, grain, coal, and oil. Oil alone had a share of 25%, much lower than its 60% share of 1970 (Rodrigue, 2017). Bulk carriers are the largest vessels currently in operation, and the largest oil tankers can reach up to 500,000 deadweight tons. Crude tankers are among the more dangerous ships for potential environmental impacts, as demonstrated by the oil spilled into the water during several accidents, the worst being Atlantic Empress in 1979, ABT Summer in 1991, Castillo de Bellver in 1983, and Amoco Cadiz in 1978 (ITOPF, 2019). Oil spills are consistently decreasing in the last five decades, with 55% occurred in the 1970s, and only 6% after 2000, thanks to increased security standards. However, it is estimated that roughly 5.9 million tonnes have been spilled from tanker accidents since 1970, which is equal to roughly 5% of the total seaborne oil, crude, and gas carried in 2017 (ITOPF, 2019), or to less than half of the daily global oil consumption in 2017 (IEA, 2018b). While this amount appears to be relatively small, it is important to remind that its local environmental impact is significantly larger than for the oil combustion, especially considering the limited area that is affected.

Ships are the less energy-intensive freight transport mode, with a specific energy consumption that is between 5 and 20 times lower than trucks on a country basis, with a world average around 0.03 kWh/tkm (2.5 toe/Mtkm) (IEA, 2019c). Two key factors impacting the fuel consumption are the ship size and the cruising speed, the latter leading to an exponential increase of energy consumption. The choice of cruising speed is generally a trade-off between the fuel costs and the duration required by the trip, which may also require the use of more ships to maintain the same frequency on port calls on an inter-range service (Rodrigue, 2017). On the other hand, the ship size has seen a significant evolution in the last decades, and currently, the maximum sizes are limited by the characteristics of major canals (mainly Panama) and ports, as well as the need of finding paying cargo to fill the ships and justify the additional investment costs.

1.4 Focus on Selected World Regions

The feature of transport systems and mobility patterns across the world shows significant variations, since the historical development of transport infrastructure and the availability of and preference for different transport modes has been driven by several aspects including geography, economy, culture, development, resource availability, geopolitics. For this reason, any future strategy dealing with sustainable mobility planning needs to be carefully designed based on local conditions, since there is no one-size-fit-all solution to deal with the different problems related to passenger and freight transport. Moreover, within each country, strong difference exists between

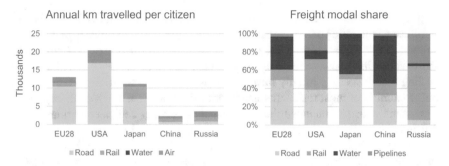

Fig. 1.9 Annual km travelled per citizen and freight modal share for selected countries. Authors' elaboration on European Union (2018)

rural and urban areas, although this dichotomy has many similarities across world regions.

Figure 1.9 presents a comparison of some indicators for selected countries, showing that significant regional differences exist both for passenger and freight modal shares. However, such comparisons should be evaluated carefully, since statistical data are often collected and aggregated with different logics, which may lead to non-comparable figures.

1.4.1 Europe

Europe includes today more than 50 countries, corresponding to more than 700 million inhabitants. The transport of passengers and goods has continuously increased since the industrial revolution, and European countries have mostly cooperated in developing and maintaining compatible transport networks. Europe is connected through a capillary network of roads, railway lines and inland waterways, as well as through marine ports and airports. The estimated transport activity in the EU-28 in 2016 reached 6.8 trillion pkm for passenger transport on any motorized vehicle (although 71% on cars), and 3.7 trillion tkm of freight transport, of which 49% on road and 32% on inland waterways (European Union, 2018).

Thanks to the significant development of multiple transport infrastructures, European countries are very well connected, and citizens can often choose from different modes for any given trip, especially in urban contexts. International trips are relatively easy, especially in the Schengen Area, which enables border control-free travel between 26 European countries. However, there are significant differences across countries, both between Northern and Southern Europe, due to cultural habits and weather conditions, as well as between Western and Eastern Europe, due to a different level of economic development. Although the European Union is developing common regulations and strategies, national policies over the years have supported different mobility paradigms.

Some Nordic countries, especially Netherlands, Denmark, and Germany, have a significant penetration of bicycle usage for commuting and other urban trips, thanks to cultural habits, dedicated policies, and geographical conformation of most cities. The key to achieving high-cycling usage in cities appears to be the provision of separate cycling facilities along heavily travelled roads and at intersections, combined with traffic calming of most residential neighborhoods (Pucher & Buehler, 2008). Other significant measures included the development of proper dedicated parking areas, full integration with public transport and promotional and educational events for citizens using bikes and cars. Bike usage in several European cities has also been supported by the development of bike sharing systems throughout the last decade, with the aim of providing the users with a first- and last-mile transport mode to be coupled with public transport to decrease the use of private cars in densely populated cities.

Most European countries have high usage rates of public transport, both at urban level and across cities. Thanks to the diffused road and rail network, trains and buses are a convenient and affordable solution for passengers needing to travel from a city to another, also between different countries. The largest European cities have underground metro lines that have been in operation since decades, and the first system in the world opened in London in 1890. As of 2017, European metro systems are present in 46 cities, carrying 10.8 billion passengers per year (UITP, 2018).

The European Union is developing climate and energy policies aiming at decreasing the carbon emissions, as well as supporting primary energy savings and the use of renewable energy. EU Directives have also set specific targets for the use of renewable fuels in transport, particularly biofuels but also electricity from renewable energy sources. In last years, the adoption of electric vehicles (EVs) is rising, although with a strong variability from a country to another. Norway is becoming the world leader for EVs, with a market share of electric cars that peaked to 46% in 2018 (IEA, 2019a), driven by the generous incentives provided by national regulations. Europe is the second marked of EVs worldwide, after China and before the USA, but with a market share that remained below 3% in 2018.

1.4.2 North America

North America is a very large continent with a relatively low population, with consequences on both passenger and freight transport modes. The USA and Canada have a large network of highways connecting the major cities, but some large rural areas are still equipped with poor-quality gravel or unpaved roads. As a result of the more recent development of this continent in comparison with Europe or Asia, many city plans have been developed based on the use of private car as the primary mobility mode, leading to lower densities and very large cities. Due to the significant distances between cities, domestic aviation has gained a strong importance in the last decades for passenger transport, especially thanks to the development of low-cost air carriers that led to an increase of air transport demand. The extended railway network is currently used almost only for freight transport, with some limited exceptions in regional areas where passenger transport is used for commuting, especially on the East Coast of the USA.

The dominance of car in the last century shaped both the society and the current US infrastructure, which would require significant investments to shift toward alternative modes such as public transport or active modes. Road transport has also been strongly supported by federal and state regulations, as well as to lower taxes on transport fuels in comparison with other developed countries. The federal support for highway projects in the US is higher than for public transit, and for passenger rail, no funding is given to the single state, resulting in significant consequences on modal competition (Rodrigue, 2017).

However, in the last years, many cities are testing innovative solutions to address the rising problems related to congestion, local pollution, and land space use in city centers. Digital technologies including sharing mobility, Mobility-as-a-Service, and autonomous vehicles are being the object of multiple startups, especially in California and other major cities across the country (see Chap. 3). The USA has seen the rise of Uber and Lyft, two ridesharing companies that reached tens of billions of market capitalization, and delivered 5 billion and 620 million of trips in 2018, respectively (Trefis Team, 2019).

Another significant trend, although with strong difference across US states, is the rise of EVs, especially in California, Hawaii, and Washington, with the USA being the second country worldwide for EVs sales in 2018, reaching 360 thousands units, 17% of the global sales (IEA, 2019a). Moreover, the US company Tesla, one of the companies at the origin of the current EVs hype, became the top selling EVs manufacturer in the world in 2018, with 245 thousand units (Demandt, 2019), mainly sold in the domestic market. However, Tesla has not yet generated a full-year profit over its 15-year history, although its performance has been getting better in recent years.

1.4.3 China and East Asia

International transport has played a crucial role in supporting the strong economic development of Asian countries in last decades, and especially in China, the construction of seaports and transport infrastructures has made possible the development of manufacturing supply chains producing goods for the global market. However, the increasing well-being and urbanization rates resulted in massive mobility demand for commuting and access to services, and since the deployment of mass transit systems has not kept the pace with the increasing demand, private cars and other motorized vehicles are creating increasing congestions and local pollution in large cities. Asian countries, especially in Southeast Asia, are also characterized by a massive use of two-wheelers and three-wheelers in cities, with significant impacts on congestions and road safety.

At a regional level, the challenge of developing transport programs that ensure inclusiveness and safety is still a major hurdle, especially for rural areas that are still lacking connections with social and economic networks (Rakhmatov et al., 2017). An improvement of rural transport is necessary to address multiple aspects including economic development, employment, access to health and education facilities, as well as poverty reduction by connecting producers and consumers. Significant differences exist from a country to another, depending on the level of development and on cultural, historical, and political issues.

As in other sectors, transport programs in China have a significant impact at a global scale, given its large population, its strong economic development as well as very large and growing urban areas. The unacceptable levels of local pollution in last years have led the government to push for electrification of passenger transport, with the largest cities deploying draconian policies to support the use of both electric cars and electric buses. Some cities are even limiting the number of fossil-powered new cars that can be sold each year, by auctioning a limited number of license plates. However, the development of EVs is also seen as a strategic action to develop the national industry, to compete with other countries that have a stronger know-how on traditional automotive based on internal combustion engines.

Similar pollution and congestion problems are being faced by other Asian countries, including India, whose population is likely to exceed the one of China, thus making India the most populated country in the world. Both road and rail networks are significantly developed and public transport remain the only option for most citizens, although private cars and two- and three-wheelers are increasing, with negative consequences on road safety. In large cities, web-based ridesharing apps are providing cheap solutions in alternative to traditional taxi cabs, providing to a larger number of citizens the possibility of moving by car on demand. However, the current infrastructure is unable to meet the strong rise in demand, and the lack of investments is slowing economic growth, especially in rural areas. Rail activity in India is expected to grow more than in any other country, with passenger movements in India reaching 40% of global activity. Rail remains the primary transport mode connecting numerous cities and regions, and passenger transport is currently second only to China (IEA, 2019c).

Transport demand in Asia is expected to show a strong increase in the following decades, both for passenger and freight, and it must face the constraints related to the international agreements on decarbonization. In particular, freight transport is supporting the increased wealth of the continent's population, but a significant share is also related to global trade. As a result, a crucial aspect in international agreements will be the shift from a production-based allocation of impacts toward a consumption-based logic, for a deployment of effective decarbonization policies (Golinucci, Rocco, & Colombo, 2019).

1.4.4 Latin America

Central and South America is characterized by a large variety of environments, with the largest cities being concentrated nearby the Eastern and Western coasts, with large rural regions in the hinterland. An efficient and integrated system of transportation is essential to foster the development of the region and the trade and passengers' movement between the different countries. The main roads network is interconnecting the largest cities of the continent, but there are huge disparities with the road conditions in rural areas, where roads are often unpaved and with a lower quality. Different projects have been developed in the past to ensure a good interconnection of the national highways, including bridges connecting Argentina, Uruguay, Paraguay, and Brazil.

The strong development of roads has caused railways to lose their dominant position after the 1960s, resulting in a decrease of the quality of the service, caused by operational problems and equipment aging. Moreover, most lines are single track, discouraging the passenger transport due to the delays, and the presence of different gauge standards is hindering an efficient interconnection between the different rail networks. The privatization of rail systems performed in the 1980s in different countries in the region has further led to a huge decrease of both passenger routed and rail freight transport (Knapp et al., 2019).

Sea transportation has been significant in the history of most South American countries, since the majority of imports and exports in the continent are relying on shipping. There are some major inland waterways, but the freight traffic is generally limited and with low potential for expansion in future years. On the contrary, domestic air transport has strongly developed in the last decades, due to the significant advantages for passenger mobility between distant cities with long and often uncomfortable roads connecting them.

Considering urban transport, South America has seen a strong development of Bus Rapid Transit (BRT), which is now present in 55 cities in Central and South America carrying more than 20.5 million passengers per day, which corresponds to a 60% share at global level (BRT Data, 2019). BRT is a system designed to improve, reliability, capacity, and speed in comparison with conventional bus systems in cities. Its main features are the use of dedicated lanes and priority at intersections, to prevent the impact of congestions on the bus speed. Other aspects to speed up the boarding times include the use of off-board fare collection, platform-level boarding, and the use of high-capacity buses.

The main rationale of BRT systems is to combine the advantages of high capacity and speed generally provided by rapid transit, with the flexibility and the lower capital cost of the bus systems. This does come at a cost, since there are usually higher operational costs and a lower lifetime of the vehicles. Moreover, although more efficient than private cars, in comparison with an electricity-powered rail transit, BRT shows higher environmental impacts both for local pollution and GHG emissions issues. Moreover, overcrowding and poor service quality are major concerns in many cities, including Santiago and Bogotà, whose TransMilenio is among the largest in the world. People are complaining about the long waiting times and the low comfort of the trips, resulting in a modal shift of a large number of users toward private cars and motorcycles in the last decade. To reverse this trend, it is important that bus operators develop new financial plans that go beyond the simple fare collection, by following the example of other cities in Europe and Asia.

1.4.5   MENA

The mobility demand in the region has rapidly increased since the 1970s, following a rise of the population, together with one of the highest urbanization rates in the world. Poor urban planning and the lack of good public transport have driven an increase of car usage, also supported by rising incomes in some areas. However, rural areas and urban citizens with lower incomes are still using nonmotorized vehicles or overcrowded minibuses (The Economist, 2016), with increasing inequality concerns in some countries.

The dramatic increase of private cars, also favored by cheap fuel prices in oil-producing countries, has not been supported by an adequate planning and deployment of infrastructures, resulting in major congestions, especially in large cities. Together with road safety issues, traffic is also significantly impacting the GDP of some countries, due to the lack of access to opportunities for the citizens and the limitations caused to trade. Moreover, the difficulties of commuting further pushe toward urbanization, leading in some cases to strong problems in providing services to the citizens in highly populated areas (The Economist, 2016).

However, congestion problems are leading to an interest in developing public transport systems in larger and richer cities, starting from the mass transit system opened in Dubai in 2009, among the longest fully automated systems in the world (Hashem, 2016). New subway systems, including the huge rapid transit system being deployed in Riyad, are exploiting the most recent technology developments to face the harsh weather conditions in the region, both for the potential intrusion of sand in the vehicles and the extreme temperatures requiring dedicated cooling systems.

Moreover, thanks to the availability of financing, cities in UAE are becoming test sites of multiple innovations in transport. In particular, Dubai aims at providing one quarter of all vehicle trips by autonomous cars by 2030, and driverless cabs are already being tested in the city. Furthermore, innovation is being pushed even farther, since autonomous electric-powered flying taxis are being considered for testing in the city. However, the innovations are not limited to urban mobility, since the connection between Dubai and Abu Dhabi is being chosen for one of the first potential applications for Hyperloop, which would allow transporting 10,000 passengers per hour between the two cities in only 12 min (Red Herring, 2018).

However, it must be kept in mind that all of these projects are still at very early stage. And while innovation is being put at the center of the transport planning agendas in few cities, much still needs to be done in smaller cities and rural areas, as well as in the poorer countries of the region. National mobility strategies should clearly aim at decreasing inequalities and promote a sustainable development for all their citizens, for which a reliable and affordable transport system is unavoidable.

1.4.6 Sub-Saharan Africa

The history of transport infrastructures in Sub-Saharan Africa had been strongly affected by the European colonial powers. While there had been highly developed transport networks in many parts of Africa in pre-colonial time, during the colonial era these infrastructures were adopted to connect seaports to the internal areas that were rich of resources, with the sole aim of serving the interests of the external powers (Kröner et al., 2019). This happened both to roads and to railways, the latter being also affected from an uncoordinated development of different gauges, hindering interconnections between different countries. All of this was further complicated by the vast unpopulated areas lying between the main centers.

This is the single region with the highest expected population increase in the following decades, which together with high urbanization rates will require reliable and effective transport systems to improve the access to opportunities and services required for higher standards of living in comparison with the current situation in the continent. While the issues of energy access and access to clean cooking are being at the center of multiple discussions, the access to opportunities supported by sustainable mobility is often underestimated. In comparison with other continents, in Africa, walking is still the most common mode of transport in most countries. And yet nonmotorized transport is not receiving the necessary attention, especially in large cities, that should become more pedestrian-friendly.

Most African cities are on a development trajectory of increasing private car usage and informal public transit, which will evolve toward unsustainable mobility patterns without a proper policy development. The demand for efficient and affordable transport systems is very high, because a large part of people's income and time is spent on their daily commute. African cities should look for leapfrog opportunities by exploiting the best practices and technology development that are being deployed in other world cities. Sustainable urban transport solutions are crucial to mitigating the growing congestion, road safety issues, and pollution in the region's sprawling urban centers (SSATP, 2018).

The region is facing a strong urbanization, with large cities all over Africa affected by congestion problems leading to higher costs. Moreover, there are still significant social, political, economic, and physical barriers to mobility that are hindering social inclusiveness. New technologies supported by an effective use of big data may play a crucial role in helping dynamic traffic management and the coordination of other resources on the road. However, the comprehensive, consistent strategies needed at the national and subnational level to tackle these challenges are lacking. Better planning, better institutional coordination, and more appropriate and sustainable financing are clearly needed (SSATP, 2018).

1.5   Conclusions and Key Take-Aways

This chapter described the main aspects related to transport, highlighting the complexity of the sector and the high variability of mobility demand and supply with respect to multiple dimensions, including geography, demography, sectors, technologies, and transport modes. Both passenger and freight transport are at the basis of an effective development of countries and societies, and the sustainability of transport is becoming more and more necessary, due to the rising concerns related to climate change, local pollution, congestions, and safety, especially in large cities all over the world. However, significant differences exist among world regions, since cultural, economic, historical, political, and geographical aspects are crucial in the development of transport modes and infrastructures.

The evolution of mobility is being shaped by two main trends, digitalization and decarbonization, which will be further discussed in Chaps. 2 and 3. Both trends are tightly related with the development of international, national, and local policies, as will be described in Chap. 4. Policy priorities may vary from an area to another, with strong effects on the support of specific transport modes and technologies, which may lead to different transport mixes in comparison with the current situation, especially due to the huge increase of demand that is expected in the following decades.

Huge differences are expected between developed and developing countries, since the lack of well-developed transport infrastructures, which is currently seen as a burden, may become an opportunity of leapfrogging toward better transport systems. Learning from the current issues of multiple transport systems in large cities worldwide, avoiding the lock-in of oversized road networks supporting mobility models solely based on non-shared private cars will be crucial, especially in urban areas.

References

Althoff, T., Sosič, R., Hicks, J. L., King, A. C., Delp, S. L., & Leskovec, J. (2017). Large-scale physical activity data reveal worldwide activity inequality. *Nature, 547*, 336. Retrieved from https://doi.org/10.1038/nature23018.

Ausubel, J. H., Marchetti, C., & Meyer, P. S. (1998). Toward green mobility: The evolution of transport. *European Review, 6*(2), 137–156. https://doi.org/10.1017/S1062798700003185.

BRT Data. (2019). *Global BRT Data—Latin America*. Retrieved August 23, 2019, from https://brtdata.org/location/latin_america.

Buehler, R., & Pucher, J. (2012, June). Walking and cycling in Western Europe and the United States. *TR News*, 34–42. Retrieved from http://onlinepubs.trb.org/onlinepubs/trnews/trnews280westerneurope.pdf.

Demandt, B. (2019, February). Global electric car sales analysis 2018. *Carsalesbase*. Retrieved from http://carsalesbase.com/global-electric-car-sales-analysis-2018/.

European Union. (2018). *Statistical Pocketbook 2018—EU Transport in figures*. Luxembourg: Publications Office of the European Union. https://doi.org/10.2832/05477.

Golinucci, N., Rocco, M., & Colombo, E. (2019). The effectiveness of LCA-based emissions policies against carbon leakage: Theory and application.

Hashem, H. (2016, December). Autonomy is shaping the Middle East's transportation future. *Think Progress*. Retrieved from http://www.think-progress.com/ae/performance-and-productivity/autonomy-is-shaping-the-middle-easts-transportation-future/.

Hyperloop One. (2019). Hyperloop One. Retrieved August 19, 2019, from https://hyperloop-one.com.

ICAO. (2018, December). Solid passenger traffic growth and moderate air cargo demand in 2018. Retrieved from https://www.icao.int/Newsroom/Pages/Solid-passenger-traffic-growth-and-moderate-air-cargo-demand-in-2018.aspx.

IEA. (2017). World Energy Balances database. In *IEA World Energy Statistics and Balances (Database)*. https://doi.org/10.1787/data-00512-en.

IEA. (2018a). *CO_2 Emissions from Fuel Combustion 2018*. Retrieved from https://webstore.iea.org/co2-emissions-from-fuel-combustion-2018.

IEA. (2018b). *World Energy Outlook 2018: The future is electrifying*. Retrieved from https://www.iea.org/workshops/world-energy-outlook-2018-the-future-is-electrifying.html.

IEA. (2019a). *Global EV Outlook 2019*. Retrieved from www.iea.org/publications/reports/globalevoutlook2019/.

IEA. (2019b). *IEA Atlas of Energy*. Retrieved August 19, 2019, from http://energyatlas.iea.org.

IEA. (2019c). *The future of rail*. Retrieved from https://webstore.iea.org/the-future-of-rail.

ITF. (2019). *ITF Transport Outlook 2019*. https://doi.org/10.1787/9789282108000-en.

ITOPF. (2019). *Oil Tanker Spill Statistics 2018*. Retrieved August 21, 2019, from https://www.itopf.org/knowledge-resources/data-statistics/statistics/.

Kasper, D. M., & Lee, D. (2017). *An assessment of competition and consumer choice in Today's U.S. airline industry*. Retrieved from http://darinlee.net/pdfs/airline_competition.pdf.

Kharina, A., & Rutherford, D. (2015). *Fuel efficiency trends for new commercial jet aircraft: 1960 to 2014*. Retrieved from https://theicct.org/sites/default/files/publications/ICCT_Aircraft-FE-Trends_20150902.pdf.

Kingsland, P. (2018). Will maglev ever become mainstream? *Railway Technology*. Retrieved from https://www.railway-technology.com/features/will-maglev-ever-become-mainstream/.

Knapp, G. W., Dorst, J. P., et al. (2019). South America. In *Encyclopaedia Britannica*. Retrieved from https://www.britannica.com/place/South-America/Transportation.

Kröner, A., Gardiner, R. K. A., et al. (2019). Africa. In *Encyclopaedia Britannica*. Retrieved from https://www.britannica.com/place/Africa/Transportation.

Litman, T. A., & Doherty, E. (2009). *Executive Summary of Transportation Cost and Benefit Analysis Techniques, Estimates and Implications*. Retrieved from http://www.vtpi.org/tca/tca00.pdf.

Maglev.net. (2018). *The six operational maglev lines in 2018*. Retrieved August 19, 2019, from https://www.maglev.net/six-operational-maglev-lines-in-2018.

Poushter, J. (2015, April). Car, bike or motorcycle? Depends on where you live. *Pew Research Center—Fact-Thank*. Retrieved from https://www.pewresearch.org/fact-tank/2015/04/16/car-bike-or-motorcycle-depends-on-where-you-live/.

Pucher, J., & Buehler, R. (2008). Making cycling irresistible: Lessons from the Netherlands, Denmark and Germany. *Transport Reviews, 28*(4), 495–528. https://doi.org/10.1080/01441640701806612.

Rakhmatov, B., Lee, C., Chong, E., Kormilitsyn, F., Ishtiaque, A., Regmi, M. D. … Tanase, V. (2017). *Review of development in transportation in Asia and the Pacific 2017*.

Red Herring. (2018, September). Progress speeds up on Dubai to Abu Dhabi UAE Hyperloop. *Red Herring*. Retrieved from https://www.redherring.com/asia/progress-speeds-up-on-dubai-to-abu-dhabi-uae-hyperloop/.

Rodrigue, J.-P. (2017). *The geography of transport systems* (4th ed.). New York: Routledge.

SpaceX. (2013). *Hyperloop Alpha*. Retrieved from https://www.spacex.com/sites/spacex/files/hyperloop_alpha-20130812.pdf.

SSATP. (2018). Africa Transport Policy Program—SSATP annual meeting 2018. In *Africa's rapid urbanization and the response to urban mobility in the digital era*. Abuja, Nigeria. Retrieved from https://www.ssatp.org/sites/ssatp/files/publications/AGMProceedings-Abuja2018_EN.pdf.

Teter, J. (2018). The future of trucks—Implications for energy & the environment. In *Stakeholder meeting on the Impact Assessment on HDV CO$_2$ emission standards*. Retrieved from https://ec. europa.eu/clima/sites/clima/files/events/docs/0121/iea_en.pdf.

The Economist. (2016, March). Let's go together. *The Economist*. Retrieved from https://www. economist.com/middle-east-and-africa/2016/03/10/lets-go-together.

Trefis Team. (2019). How do Uber And Lyft compare in terms of key revenue and valuation metrics? *Forbes*. Retrieved from https://www.forbes.com/sites/greatspeculations/2019/04/22/how-do-uber-and-lyft-compare-in-terms-of-key-revenue-and-valuation-metrics/.

UITP. (2018). *World metro figures 2018, 8*. Retrieved from https://www.uitp.org/sites/default/files/ cck-focus-papers-files/StatisticsBrief-Worldmetrofigures2018V4_WEB.pdf.

United Nations. (2018). *World urbanization prospects: The 2018 revision*. Retrieved from https:// esa.un.org/unpd/wup/.

WHO. (2011). *Health in the green economy—Transport sector*. Retrieved from http://www.who. int/hia/examples/trspt_comms/hge_transport_lowresdurban_30_11_2011.pdf.

WHO. (2018). *Global status report on road safety 2018*. Retrieved from https://www.who.int/ violence_injury_prevention/road_safety_status/2018/en/.

World Bank. (2019). *Databank*. Retrieved September 24, 2019, from https://databank. worldbank.org.

Chapter 2
Decarbonization Solutions

Abstract This chapter will be devoted to a description of the alternative decarbonization pathways that may help to decrease the dependency from fossil fuels in transport, which is currently the sector most strongly dependent on oil. Electricity, hydrogen, and biofuels are the main alternative sources for transport systems, and they will be analyzed and compared by considering the state of the art of the technologies of each pathway and the potential future development. Each transport segment and each mode have specific features, resulting in the need to evaluate dedicated applications based on the technical and economic conditions of each technological solution. Moreover, variable conditions across world regions may impact the sustainability of each pathway, particularly in relation to the current and expected power generation mix that varies from a country to another. Opportunities and challenges will be discussed to provide to the readers a clear vision on the strengths and weaknesses of each solution.

2.1 Introduction

The continuous increase of carbon emissions related to human activities is leading to rising greenhouse gases concentrations in the atmosphere, causing an increase of average world temperature. Since 1992, an international effort to "stabilize greenhouse gas concentrations in the atmosphere at a level that would prevent dangerous anthropogenic interference with the climate system" has been declared by the United Nations Framework Convention on Climate Change (UNFCCC). In this framework, the Paris Agreement signed in 2016 aims to keep the increase in global average temperature to well below 2 °C above pre-industrial levels. Although these efforts may not be enough to effectively tackling the risks of climate change, there seems to be an increasing awareness both in governments and in public opinion of the importance of decreasing carbon emissions. Global energy-related carbon emissions peaked to an historic high of 33.1 Gt of CO_2 in 2018, with a 1.7% rise over the previous year (IEA, 2019a).

Around one quarter of the total emissions are related to transport, which remains heavily dependent on oil products (accounting for more than 90% of its final energy

© The Author(s) 2020

M. Noussan et al., *The Future of Transport Between Digitalization and Decarbonization*, SpringerBriefs in Energy, https://doi.org/10.1007/978-3-030-37966-7_2

consumption). Moreover, transport is among the sectors that are most difficult to decarbonize, especially for some segments including international shipping and long-haul flights. Different technologies may play a role in transport decarbonization, including renewable electricity, green hydrogen, biofuels, and synthetic fuels. While some solutions are already mature, their costs are often higher than the traditional oil-based technologies, and dedicated policies may be required to trigger a decrease of costs driven by economies of scale.

Carbon emissions generated by the transport of passengers or goods show a high variation depending on the mode, the technology, the age of the vehicle, the fuel, the number of passengers or the load factor, the driving cycle, etc. Table 2.1 reports some average values and ranges of variation for different countries, to compare the effectiveness of various transport modes. Since all the aspects mentioned above have an impact on emissions, it is important to highlight that these average values should only be considered as a reference for current transport modes, but they are significantly dependent on the vehicle fleets and the way people use them.

Today, CO_2 emissions related to the transport sector are around 8.2 Gt, including the emissions caused by the generation of the electricity consumed in transport (IEA, 2019a). The future evolution of these emissions will result from the combination of a strong expected increase of the demand of mobility of both people and goods, a foreseeable increase in energy efficiency of different transport modes (including the potential for higher load factors, especially for cars), and a possible shift toward low-carbon energy sources for vehicles. Based on the most up-to-date scenarios published by the International Energy Agency (IEA, 2019f), the CO_2 emissions caused by transport may increase up to 10.2–11.5 Gt by 2040 (authors' calculation

Table 2.1 Average specific CO_2 emissions for different transport modes and variation ranges

	Average	Range of variation[a]
Passenger (gCO₂/pkm)		
Large cars	196	74–262
Aviation	130	75–221
Cars	130	60–221
Buses and minibuses	44	26–85
Two- and three-wheelers	31	24–54
Rail	18	6–81
Freight (gCO₂/tkm)		
Medium trucks	96	48–152
Heavy trucks	80	58–117
Rail	15	9–52
Shipping	8	6–13

Authors' calculation from IEA (2019e)
[a]The variation is to be intended on a country basis, while stronger variations may be expected when considering a single vehicle or trip

based on the outcomes of Stated Policies and Current Policies scenarios) or they may decrease to 5.8 Gt in the best case (Sustainable Development Scenario). Transport will remain responsible of 28–37% of the total emissions in 2040, and it will remain among the most difficult sectors to decarbonize, together with heavy industry.

While there is a strong policy push toward decarbonization in multiple countries worldwide, low-carbon transport technologies alone are not the solution, but they should come together with significant energy efficiency measures, multimodal transport optimization, and mobility demand control. Sustainable mobility planning strategies with the integration of different governance levels are unavoidable, to ensure a development of the sector characterized by lower impacts related to both energy consumption and carbon emissions. Reaching the challenging decarbonization targets that are required to limit climate change may need a combination of different solutions, since there is no single silver bullet to supply all the mobility demand related to passenger and freight transport at different scales and with different modes.

2.2 Electricity—The Main Option

Electricity is already seen as a clean energy source due to the absence of local pollution emissions, although its generation in power plants may include the use of fossil fuels combustion, with consequent centralized pollutants emissions, which are usually monitored and limited by dedicated abatement systems. However, as long as GHG emissions are concerned, from a well-to-wheel (WTW) perspective, i.e., considering the entire energy supply chain, the use of electricity may not guarantee a lower impact than traditional oil-based fuels such as gasoline and diesel oil.

For this reason, the push toward electrification of transport and other final uses needs to come along with a decarbonization of the power generation mix, by supporting low-carbon sources such as renewables or nuclear, or alternatively by coupling fossil-powered plants with carbon capture, utilization, and storage (CCUS) systems. A strong penetration of non-dispatchable renewable energy sources (RES), such as solar or wind, may require the support of proper flexibility solutions to match energy demand and supply. Although the discussion of such aspects goes beyond the scope of this book, some insights will be given when dealing with specific technologies, and the reader may find additional information on dedicated literature works (Koltsaklis, Dagoumas, & Panapakidis, 2017; Sinsel, Riemke, & Hoffmann, 2020).

At global scale, the carbon emissions caused by the generation of a unit of electricity are showing a slightly decreasing trend, and currently, the global average carbon emissions intensity is equal to 518 g CO_2-eq/kWh (IEA, 2019b), but large differences remain across countries. This decrease is happening both thanks to an increase of conversion efficiency and a growing share of RES in the generation mix, although new fossil-based capacity is still being installed worldwide. Figure 2.1 reports the historical evolution of the national carbon intensity of electricity generation for the ten largest countries for power production in 2015 (most recent data available). As

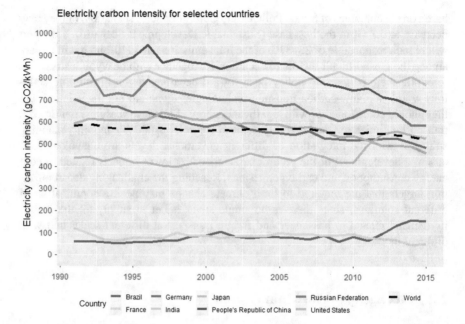

Fig. 2.1 Carbon intensity for electricity generation in selected countries. *Source* Authors' elaboration from IEA (2019c)

already noticed, strong differences exist, and while a decrease is noticeable for China and the USA, most countries have a relatively constant carbon intensity, and Brazil and Japan have even had an increase in the very last years on record.

However, the specific emissions are only one side of the story, since the total emissions are consistently rising, driven by a strong expansion of electricity demand worldwide, due to the electrification of different final uses and to the increasing electricity access rates and consumption in developing countries. It is not clear if the increasing deployment of RES plants will keep the pace with the rising demand without a supportive regulatory framework, and therefore, dedicated decarbonization policies are of utmost importance to limit the current trend of climate change.

Electricity has been widely adopted for a wide range of applications, being a flexible, efficient, and clean energy vector. However, one of the toughest barriers to electricity use is the difficulty of storing it with efficient and inexpensive solutions, since supply and demand need a continuous matching to ensure the stability of the power networks. Electricity storage has wide applications at small and medium sizes, from portable electronics to some appliances, but medium- and large-scale batteries have not yet been proven to be economically affordable nor environmentally sustainable, mainly due to the high energy consumption and impacts in the manufacturing and decommissioning stages. Currently, the only option available for large-scale electricity storage is pumped hydropower, which is however limited to sites with proper geographical conformation.

Nevertheless, electric transport is seeing a considerable boost in last years, supported by the double aim of decarbonizing the transport sector and decreasing the severe local pollution that is affecting most urban centers worldwide. The following sections will present the characteristics of the main technologies available for different transport modes, as well as the aspects related to the energy supply chain and infrastructure, together with some insights on current case studies that reflect some highlights of this opportunity.

2.2.1 Electric Transport Technologies

Electricity has already a long history in transporting passengers and goods, notably in electrified railways that can power both extra-urban trains and urban rapid mass transit systems underground or at ground level. In some cases, network-powered trolley buses have been used for urban public transport. However, all these systems required to be constantly connected to a power source, resulting in higher investment costs and a limited flexibility. Although these systems are still successful and gaining interest worldwide, the current electrification trend is oriented toward the use of electric batteries to store the energy required by the vehicle and provide the user with the flexibility that is usually available with oil-based cars or trucks.

The use of battery electric vehicles (BEVs) is still at an early stage, but some countries are developing policies and regulations to boost the adoption of these technologies, especially for light-duty vehicles (LDVs). BEVs are often providing limited ranges, and given the low availability of charging infrastructure in many countries, together with the longer charging times in comparison with oil-products refueling stations, the rate of adoption is still limited. On the other hand, the market share of plug-in hybrid vehicles is already reaching interesting values and different countries, thanks to the possibility of exploiting the electricity use where it is available and relying on a safe gasoline backup in the other cases.

The electric car sales are expanding at a rapid pace, and in 2018, the global electric car fleet, considering both BEVs and plug-in hybrids, reached a record of 5.1 million, up 2 million from the previous year (IEA, 2019b). At this stage, specific policies are critical in supporting the development of EVs market, leading to large differences across countries. The positive response of the private sector to these policy signals is encouraging, and the positive momentum for electric mobility is confirmed by increasingly ambitious targets from automotive manufacturers, followed by boosting investments in the battery development as well as in the charging infrastructure.

Electric cars provide several advantages, ranging from more efficient and silent engines, to no tailpipe emissions and a better driving experience and performance. The use of regenerative braking allows recovering energy during the deceleration phases, increasing the overall energy balance of the vehicle. However, the need of carrying the batteries, which are still lacking in both volumetric and gravimetric energy density, increases the weight of the vehicle, and the higher resistance required by all the structural components is adding up additional weight. Moreover, the very high efficiency of the electric engine leaves limited energy to be dissipated as heat, and this aspect becomes critical in cold climates, since the dissipation heat traditionally used in cars to guarantee an acceptable indoor temperature is no longer available. As a result, the energy consumption of BEVs in cold climates may strongly increase, leading to lower ranges and additional costs in comparison with nominal conditions (Liu, Wang, Yamamoto, & Morikawa, 2018). Ambient cooling in hot climates is also impacting the specific energy consumption, but just as it happens in fossil-powered vehicles, although this should be considered when evaluating consumption figures.

The comparison between the life cycle energy consumption of an electric car with a traditional gasoline car is mostly dependent on the electricity mix. The largest impact of traditional cars happens during the operation phase, caused by fuel combustion, while electric cars have a larger impact during the manufacturing and end-of-life phases, mainly caused by the batteries. With the current world average electricity mix, both BEVs and plug-in hybrid show lower carbon emissions over their lifetime in comparison with an average traditional gasoline car (IEA, 2019b). However, while in countries with low-carbon electricity mixes the emissions savings are considerable, traditional cars still remain a better alternative to EVs in power systems dominated by coal generation, as long as carbon emissions are concerned. This aspect reflects the importance of developing policies that support EVs deployment together with low-carbon electricity sources, since EVs alone are not enough to ensure the decarbonization of the energy system.

The limited range of the vehicles, together with the weight of the batteries, becomes even more impactful for buses and trucks. In some cases, the autonomy of urban buses is currently lower than the one that would be required by their daily schedule, resulting in the need of using more vehicles to guarantee the same level of service of traditional diesel buses (Levy, 2019). This adds up to the already higher cost of electric buses, together with the charging infrastructure. Electric trucks have similar concerns, although their operation usually deals with longer travels, usually across countries, resulting in the need of additional flexibility. The current limited availability of charging infrastructure is hindering the development of such technologies, but some manufacturers are already testing electric trucks with acceptable ranges (Hanley, 2019).

An alternative solution for freight road transport electrification is the installation of catenaries in some highway networks (Barnstone & Barnstone, 2018), to be coupled with full-electric trucks with smaller batteries, or with the capacity to recharge during the travel. An alternative option is the possibility of using hybrid trucks, which can operate with higher flexibility thanks to the backup generation with traditional fuels allowing longer ranges outside of the electricity-powered infrastructure.

The challenges related to the electrification of land transport increase their magnitude when considering the potential of electrification in shipping and aviation. These transport modes are mainly relevant for passenger and freight transport over long distances, which is not compatible with the potential ranges of the current batteries nor with the reasonable expectations for new battery developments. Moreover, the charging during the travel is obviously not an option, and therefore, the electrification potential is limited to short- and medium-haul flights (although with strong challenges) and to coastal and inland navigation for water transport.

Short-haul flights may benefit from an indirect electrification: a modal shift toward high-speed trains in some national and international connections. This trend is already happening in key routes in Asia and Europe, where high-speed trains are already providing comparable performances in both prices and speed (Bachman, Fan, & Cannon, 2018). The current trade-off travel distance is around 1000 km, but future policies aiming at decarbonization may further increase the profitability of electricity-driven high-speed trains and increase this distance. Moreover, future technologies such as Hyperloop may provide additional alternatives to planes: a magnetic levitation train operated in an evacuated tube to reduce air resistance may be operated at very high speeds with little energy consumption. However, the technology is still at a very early stage, and effective demonstration systems are needed to verify the potential benefits for transport systems.

2.2.2   Electricity Supply Chain and Infrastructure

The penetration of electric vehicles involves additional aspects related to the available infrastructure, since the installation of charging stations will have an impact on the electricity supply chain. The energy consumption of electric vehicles is usually much lower than for oil-based vehicles, thanks to the higher efficiency of electric engines. However, electricity needs to be generated, distributed, and stored to be available to recharge vehicles when they need it.

Let us start from the end of the chain: the charging station. Considering electric cars, the basic possibility is to recharge it overnight using the existing low-voltage network at home or during the day at the office. A quicker option is given by chargers installed in parking lots, which are usually equipped with higher output power and allow the users to partially or fully charge their car while parked, generally at a higher price that for residential charging. A third option is given by fast-charging stations: with installed powers higher than 40 kW (and up to 150 kW in some cases), they can provide very fast recharging solutions, especially needed for long trips. While lower recharging times are more attractive for the users, although the higher economic price may be a deterrent, the impact on the power network is more challenging, since the power grid needs to guarantee a simultaneous balance between generation and demand. High penetration of fast-charging stations may require additional investments to guarantee the grid stability, including dedicated measures for distribution grids and flexibility options, including demand response and electricity storage systems.

However, while fast charging may lead to issues in specific days of high demand (e.g., when leaving for holidays), the large majority of the trips are related to short and medium distances, and thus, EVs can be recharged at lower speeds when they are not in use. The efficiency of the power system can be maximized by performing smart charging strategies, i.e., by timing the electricity supply to vehicles based on available generation from low-carbon sources. However, this results in the need of having each vehicle connected to the network for a longer time, causing a larger number of charging points for the same electric fleet. A further evolution may be the possibility of using each vehicle as an energy storage system, not only by delaying charging but also as an electricity source to provide additional flexibility to the grid when needed (often called vehicle-to-grid, V2G). However, this possibility requires that the vehicles are equipped with specific tools to allow a bi-directional power flow (which is rarely the case in current vehicles), and there are some doubts on the potential faster degradation of the battery over time. Moreover, the economic remuneration of flexibility services may reach very high values to be attractive for electric cars' users. The installation of charging stations is facing a chicken-and-egg issue during the early adoption of electric vehicles. While customers are afraid of choosing an electric car due to the lack of a proper charging infrastructure, especially for long trips, private and public investors are reluctant to deploy charging stations for the fear of overestimating the future evolution of electric cars fleets.

While charging stations are becoming the standard for electric cars power supply, other solutions are being proposed to face the problems related to range anxiety and to the duration of the charging process. A totally different model is relying on battery swapping, i.e., the idea of substituting depleted batteries instead of charging them. This model had been used in the very beginning of electric vehicles history, at the beginning of 1900 in the USA. The main advantage is related to the very quick operation, since the swap process can require even less than a gasoline refill (Tesla managed to complete a swap operation on a Model S in just 90 s during a test event), and without the need for the driver to exit the car. Moreover, the idea of a third-party company which owns and operate the batteries, selling the service to the users at a reasonable price, may decrease the cost of the electric cars making them competitive with the traditional ones. However, this new business model may require a very high degree of battery standardization, eventually over different car models, which appears rather difficult in the current market. An additional advantage of a significant number of batteries in a swap station may be its potential to participate to V2G programs. Some pilot plants have been in operation in the last decade in different countries, and although some of them have been successful especially for scooters and electric bikes, none of them has yet proven to be an effective alternative to charging stations for electric cars. However, some battery swapping programs are being evaluated in China, with particular focus on specific fleets, such as municipal taxis or buses.

2.2.3   The EV Momentum in Different World Regions

Electric vehicles, especially passenger cars, are showing increasing penetration rates, and the first three markets in 2018 were China, Europe, and the USA, with car stocks at the end of the year reaching 2.3, 1.2, and 1.1 million vehicles, respectively (IEA, 2019b). However, the market share of new electric vehicles sales remained under 5% in all countries but four (Norway 46%, Iceland 17%, Sweden 8%, Netherlands 7%). While electric light-duty vehicles show significant increase year-over-year, the large majority of electric vehicles worldwide remain two/three-wheelers: they exceeded 300 million at the end of 2018, the large majority being in China. Two-wheelers still account for the largest share of the 58 TWh of estimated consumption of global electric vehicles fleet in 2018, of which 80% is related to China. On a well-to-wheel basis, the (IEA, 2019b) estimates a total emission of the global EV stock in 2018 at 38 Mt of $CO_{2\text{-eq}}$, to be compared with 78 Mt $CO_{2\text{-eq}}$ emissions that an equivalent internal combustion engine fleet would have emitted.

The adoption of EVs is strongly driven by dedicated policies that aim at supporting low-carbon solutions as well as decreasing the tailpipe emissions in urban environments. This latter target is especially important in Chinese cities, which have implemented strong regulations in different sectors to limit the pollutants emissions. The central government of the country has proven to be effective in deploying large fleets of EVs, including a significant number of electric buses in several cities. China accounts for 99% of the global electric buses stock, which reached around 460,000 units at the end of 2018. The country had also pursued an industrial strategy aiming at becoming the world leader in manufacturing electric vehicles, especially on the batteries and the supply chains of the required materials (investing on refining plants for lithium, cobalt, and graphite).

Other countries are generally supporting EVs as a solution to lower their carbon emissions, from European Union to South Korea, Japan, New Zealand, and some states of the USA, where California is accounting for the large majority of EV sales in the country. In the USA, policy packages are being developed by states, cities, and utilities, with different measures related both to incentives for EV purchase and operation and to deploy networks of charging points at workplaces or other public places. In 2018, the higher success of EVs in the country has been on the West Coast, with San Jose reaching 21% of EVs market share, and other major cities in California, together with Seattle and Portland, in the range 4–13% (Slowik & Lutsey, 2019).

In Europe, the deployment of electric vehicles shows large differences from a country to another, due to the different policies that are set in place to reach the broader decarbonization targets, especially in the member states of the European Union. Norway represents by far the world country with the highest market share for EVs, thanks to generous incentives for the switch to low-emission cars for their citizens. The country can exploit one of the most renewable electricity mixes in the world, with 95% of the total generation from hydropower and 2.6% from wind power in 2018 (Statistics Norway, 2019). The monthly EV market share has surpassed 50% for the first time in March (reaching 58.4%), and the forecast for 2019 is to remain near 50% as an annual figure. The parliament has voted an aim stating that by 2025 only zero-emissions cars (powered by electricity or hydrogen) should be sold in the country. In parallel, while the market is becoming more mature, the generous incentives are being gradually removed (although the main ones will stand till the end of 2021), and this aspect may slow down EV adoption for future users. While strong policy actions are proving to be effective in Norway, some of them may have rebound effects in the effort of promoting a modal shift from private cars to public transport, including the free access to bus lanes and the free parking for electric cars.

Moreover, Norway is a small country with a low population density and a very high average income, and thus, this success story may be harder to replicate in countries with different conditions.

2.2.4   Hydrogen—An Alternative or a Complement?

Vehicles powered by hydrogen may become a promising alternative to battery electric vehicles, thanks to the possibility of providing longer ranges and shorter refueling times. At the same time, they can become a complement to electric vehicles by addressing specific transport segments for which electrification may not prove to be effective. However, hydrogen technologies have still higher costs, especially considering the supply chain to produce, store, and transport this energy carrier. Just like electricity, hydrogen needs to be generated, and while it holds the potential of being produced by low-carbon sources, the current world hydrogen demand is almost totally fulfilled by using either natural gas or coal as primary source. Low-carbon alternatives include the use of electrolysis supplied by low-carbon electricity or coupling the current fossil-based solutions with effective carbon capture utilization and storage (CCUS) technologies. The success of hydrogen will depend of the advantages against electric vehicles, especially in sectors that are harder to electrify (such as trucks and ships), but there is the need of a significant cost decrease for the different steps of the supply chain and the transport technologies that are involved. Moreover, the implementation of an effective network of refueling stations may significantly increase the initial investment, slowing down the adoption of vehicles, especially for private users.

2.2.5   Transport Technologies Based on Hydrogen

Hydrogen-based powertrains may have an important role in different segments, including passenger cars, trucks, buses, and rail (for non-electrified lines). Hydrogen vehicles are lighter in comparison with electric ones thanks to the absence of a large battery, and they can generally provide longer ranges and shorter refueling times. They are particularly attractive for high-duty vehicles, such as trucks and buses, that would require very large batteries for their operation. Hydrogen vehicles are powered by fuel cells, which are the most expensive component, coupled to electric engines. The hydrogen is stored as a compressed gas at very high pressures (350 bar in HDVs and 700 bar in LDVs), which may lead to potential safety risks and thus require dedicated procedures and infrastructures.

Figure 2.2 reports a timeline of the expected commercial availability of hydrogen technologies in different transport segments, proposed by (Hydrogen Council, 2017). The chart highlights both the start of commercialization and the mass market acceptability, showing the different time required to penetrate each market. While

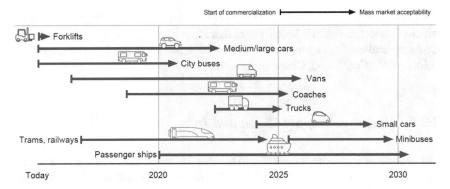

Fig. 2.2 Expected commercial availability of selected hydrogen transport technologies. Authors' elaboration on Hydrogen Council (2017)

some technologies are already available, most of the others are expected to appear in the next five to ten years, although their penetration will likely be related to external factors including an effective deployment of a clean hydrogen supply chain and a competitive price with other powertrains.

Besides passenger cars, hydrogen-powered fuel cells are already being considered for other transport segments. Hydrogen buses are being tested in different locations, in particular in Europe, Japan, South Korea, and China, with fuel cells supporting larger buses over longer distances and with fewer interruptions than electric battery solutions. Coaches and intercity buses travelling for long distances are among the most profitable applications for hydrogen powertrains. The lack of infrastructure is less problematic, since municipal buses often rely on dedicated refueling stations. An even larger potential lies in trucks, especially considering the expected increase in road freight transport in the next decades. Some manufacturers are already testing hydrogen trucks (Toyota, Hyundai, and Nikola) that should be commercially available in the next few years, but parallel investments in hydrogen supply infrastructure will be fundamental. Also, Chinese firms and municipalities are investing in hydrogen trucks and buses, with the aim of decreasing their cost thanks to economies of scale (Liu, Kendall, & Yan, 2019).

Other applications include the substitution of diesel-powered trains on non-electrified tracks, which are already being tested in Germany and China, and the passenger ships such as river boats, ferries, and cruise ships. The main driver for the success of recreational activities will be the lower local emissions, water pollution as well as the decreased noise. Some projects are already under development in Germany, France, and Norway. Finally, more than 15,000 hydrogen-powered forklifts are already operating in warehouses today, including large projects in the USA by Amazon and Wallmart (Hydrogen Council, 2017), thanks to their lower costs than electric solutions when high uptime is required.

To give some numbers on hydrogen-powered road transport, as of the end of 2018, the global hydrogen vehicles stock exceeded 12,900 vehicles, with a 80% increase over the previous year (AFC TCP, 2019). Almost half of the vehicles (46%) are in the

USA, followed by Japan (23%) and China (14%), the latter being the only country in which commercial vehicles have a larger share than passenger cars. Several countries are fixing challenging targets for the next decades, including South Korea, Japan, the USA (mainly California), China, and Europe.

2.2.6 Hydrogen Supply Chain and Infrastructure

The success of hydrogen transport will require the development of a proper infrastructure or refueling stations, together with a supply chain that is able to fulfill the demand by relying on low-carbon sources. This will require significant investments, and just like for electric charging stations with EVs, this infrastructure will probably need to be built before hydrogen vehicles reach interesting diffusion. This will likely require investors to wait a long time before seeing eventual returns, and while EVs could also benefit from existing infrastructure (e.g., home chargers), today's hydrogen demand is mostly concentrated in industrial facilities. In an early stage of development, transport applications with regular schedules (e.g., buses) may allow higher utilization rates for refueling stations, in the case of a careful planning and sizing of the expected demand.

The deployment of filling stations is a fundamental aspect to support a wide adoption of hydrogen vehicles in different segments. By the end of 2018, around 375 refueling stations were in operation worldwide (AFC TCP, 2019), with the three countries with most publicly available stations being Japan (100), Germany (60, and the USA (44). National plans to support the deployment of refueling stations aim at reaching 1000 stations by 2030 (in China and Germany) or by 2028 in France, while South Korea aims at 1200 stations by 2040. In a first phase, the development of hydrogen trucks may require fewer stations than for passenger cars, thanks to the limited number of routes and the more regular frequency of operation (Heid, Linder, Orthofer, & Wilthaner, 2017).

In parallel to the final supply of hydrogen to vehicles, an upscale of the current hydrogen generation capacity from low-carbon sources will be necessary to deliver the benefits related to the shift from current oil-based technologies. While electrolyzers are already mature technologies, their cost is still higher than hydrogen production from fossil fuels (mostly natural gas and coal). While coupling natural gas with CCUS may prove to be less expensive, attention must be paid on methane losses to avoid rebound effects. On the other hand, current electrolyzers have generally high investment costs, which need to be compensated by high utilization factors, which is not always the case when using dedicated renewable sources such as solar or wind power. An additional key parameter for the competitivity of electrolyzers is the electricity cost. Any technology will need to be able to deliver clean hydrogen at an acceptable cost, either green hydrogen (from RES) or blue hydrogen (from fossil coupled with CCUS), to support competitive solutions for different transport segments.

An additional cost in the supply chain is related to storage and transport of hydrogen, since it will not be possible to generate where and when it is required by the user. The two main solutions are to store it as a compressed gas or either in liquid form, but while the former requires pressures up to 900 bar, the latter requires temperatures as low as 20 K (which means—253 °C). Both processes—compression and liquefaction—require a significant amount of additional energy and dedicated facilities, resulting in additional costs. The choice of the best solution should be done on a case-by-case basis, due to the multiple parameters involved, and the main aspects are usually the time span of the storage (liquid hydrogen involves a fraction of losses due to boil-off) and the distance over which it will need to be transported.

For short transport distances, the main alternatives include trucks, both for compressed or liquefied hydrogen, and pipelines. In some cases, existing natural gas assets may be directly used for hydrogen, if some technical requirements on the quality of material are met. Some countries are already testing a blend of hydrogen and natural gas, but this would not be an acceptable solution for vehicles powered by fuel cells, unless pure hydrogen could be separated again from natural gas. For long-haul international maritime transport, some companies are focusing on liquid hydrogen, although other solutions include the synthesis of other chemical components that can be kept almost at ambient temperature and pressure in a liquid form, such as ammonia or other liquid organic hydrogen carriers (LCOH). In this latter case, an additional process is needed to reconvert these chemicals to pure hydrogen, unless ammonia can be used directly (e.g., as a propellent for ships, either with internal combustion engines or fuel cells).

2.2.7 Case Studies and Applications

While in different places around the world, hydrogen has already been used for testing different transport systems, some countries are now developing broader policy strategies to support a coherent and wider development of hydrogen use in their energy systems. A large vision based on multiple sectors is essential in order to guarantee that the hydrogen demand would be matched by a supply of clean hydrogen, to ensure lower carbon emissions and pollution in comparison with the traditional oil-based technologies.

The first country ever to envision the possibility of developing a hydrogen economy has been Iceland. Back in 1998, the government signed an official document proclaiming its intention to reduce Iceland's fossil fuel consumption in transport and fishing by developing hydrogen-based alternatives. In 1999, Icelandic New Energy was formed, a public–private company with the aim of supporting the transition toward a Hydrogen Society by 2050. The company was a joint venture of Daimler, Shell, Norsk Hydro, Iceland National Power Co., Reykjavik Energy, and the University of Iceland (ClimateWire, 2009). Iceland was already producing hydrogen from electrolysis for fertilizers, thanks to the abundance of clean and cheap electricity from hydropower and geothermal resources. The annual production of 2000 tons of hydrogen would have been scaled up to 80–90,000 to supply the entire transportation and fishing sectors, reducing by 66% the carbon emissions of the country. In the following years, the company tested three hydrogen buses in Reykjavik and a commercial hydrogen refueling station, and 16 passenger vehicles were in operation in the capital. But after an interesting start, this experiment has strongly slowed down and finally stopped, also because of the strong economic turmoil that invested the country in 2008. Other problems were related to the low interest from manufacturers to produce hydrogen cars, since Iceland was a limited market and there was no significant interest from other countries, resulting in high costs related to limited production scale. Today, Iceland is still considering the possibility of using hydrogen, but in parallel to other available technologies, and Icelanders recently showed a great interest in electric vehicles, which are an effective alternative to fossil fuels imports, given the abundance of clean electricity generation on the island.

Today, another country that is showing a significant interest in hydrogen is Japan, which is committed to pioneer the world's first "Hydrogen Society," through a Basic Hydrogen Strategy released at the end of 2017. The country is aiming at becoming the world leader in innovative hydrogen technologies, with the aim of developing overseas hydrogen supply chains to diversify its primary energy supply, which is currently mostly relying on imported fossil fuels. Moreover, Japan considers the development of an international hydrogen supply chain as a significant opportunity to decrease the carbon emissions of multiple sectors, including power generation, industry, and mobility. Japan's state-backed approach is ambitious, as it involves domestic and overseas industry and government stakeholders on several cross-sectoral pilot projects (Nagashima, 2018), and international cooperation will be a crucial step for the success of the entire strategy. As long as transport is concerned, the country has set increasing targets for fuel cell vehicles, up to 800,000 units by 2030, which compares with a current passenger cars stock in the country of around 69 million (Statistics Japan, 2019). Other goals are set on hydrogen buses, trucks, and forklifts.

Other countries are declaring interest for hydrogen at different levels. Australia has recently published a "National Hydrogen Strategy" (with official translations in Korean and Japanese), with the aim of positioning the national industry among the world leaders by 2030 (COAG Energy Council, 2019). Germany is currently working on a hydrogen strategy, to support the role of national industries and utilities in securing a global leadership in hydrogen technologies. Other countries are gradually showing interest and testing applications in different sectors, and transport is among the most promising.

2.3 Biofuels—A Possible Complement?

Biofuels represent today the largest share of low-carbon energy use in the transport sector, and their production has seen an increasing trend in the last decade, thanks to dedicated supporting policies, especially in the European Union, in the USA, and in Brazil. Biofuels are usually blended with traditional oil products for road transport, and thus they do not need a dedicated distribution infrastructure, since they can exploit the existing network of gasoline and diesel refueling stations. Concerns have been raised in the last years about the sustainability of biofuel production, considering the energy consumption during the production and transportation phases, as well as the effect of direct and indirect land use changes. For this reason, new policies are focusing on supporting advanced biofuels, whose feedstocks are not in competition with products used for food or feed.

While the role of biofuels blends is expected to gradually decrease due to the phase out of oil products in the long-term, in the short- and medium-term, they represent a very important tool to decarbonize transport. Moreover, bioenergy may remain the easier solution to decarbonize specific sectors such as aviation and international shipping, since alternative technologies are not yet matching the specific requirements in terms of fuel energy density. On a global basis, the IEA Sustainable Development Scenario estimates that biofuel production needs to triple to 280 Mtoe per year by 2030, representing 10% of final transport demand, compared to 3.5% today (IEA, 2019f).

2.3.1 Liquid Biofuels—Conventional and Advanced

The distinction between conventional and advanced biofuels, often referred as first and second generation, respectively, lies in the type of feedstock required for their production. While conventional biofuels rely on crops that are often in competition with production for food or feed, the aim of advanced biofuels is to exploit residues and wastes (including vegetable oils or animal fats) or energy crops that are cultivated on less productive and marginal areas, thus with a lower probability of land-use change impacts. Currently, the advanced biofuels represent only 1% of the total liquid biofuel production at global level, due to the higher costs caused by less mature technologies and supply chains, but their development is being supported by different policies in the main biofuel markets, including Europe, the USA, and Brazil (IRENA, 2019).

Conventional biofuels are generally divided into ethanol and biodiesel, which are blended at different levels with gasoline and petroleum diesel respectively. Global ethanol production reached 61.5 Mtoe in 2018, representing roughly two-thirds of total biofuel production for transportation (IEA, 2019d). The largest producer remains in the USA, which produced more than half of the world's output (33.5 Mtoe) using mostly corn as feedstock. Ethanol generation from sugarcane in Brazil reached a record of 18 Mtoe in 2018, while the rest of the world totaled 10 Mtoe only (mainly in EU, China and India). Ethanol is mainly blended to gasoline in different world countries, to reach policy mandates that are often set on specific blending targets. The share of ethanol in the fuel supplied to final users is generally limited to 5% (named E5) to 10% (E10) in Europe and the USA, to maximize the compatibility for existing vehicles running on gasoline, but in Brazil the proportion of ethanol can be as high as 25%. In the USA, ethanol is also available as E85, called also flex fuel, but it can be used only on vehicles that are specially designed to be operated with it (which total to around 20 million vehicles in the USA, i.e., 8% of the total fleet), and the current infrastructure of refueling stations remains limited. Flex fuels vehicles represent the majority of vehicles sales in Brazil, while it remains marginal in Europe, although E85 is widely available in Sweden, France, and Germany.

In addition to ethanol, one-third of transport biofuels is represented by biodiesel and hydrogenated vegetable oil (HVO), the latter remaining limited to a marginal share. Biodiesel can be extracted by different oilseed crops, although the most popular feedstocks are rapeseed in Europe, soybean in the USA and Brazil, while Asian countries generally use palm, coconut, and jathropa oils. Biodiesel and HVO production reached a global 33 Mtoe in 2018, mostly in Europe (11.6 Mtoe), the USA (6.1 Mtoe), and Brazil (4.0 Mtoe). As for ethanol, biodiesel is usually blended with petroleum diesel, with shares limited to 5% (B5) or 20% (B20) without noticeable differences with traditional oil-based diesel fuel.

Among advanced biofuels, HVO is the only one that is showing an interesting potential, although it still remains marginal in comparison with traditional biodiesel. However, the hydrogenation process that is used for its generation has the advantage of producing a fuel with very high quality, potentially better than the equivalent oil-based diesel. Moreover, the output can be further processed to produce bio-jet, a very interesting solution to decarbonize the growing energy demand of aviation. In addition to HVO, other advanced biofuels include ethanol from lignocellulosic biomass, sustainable sourced biodiesel, and various drop-in fuels refined through thermochemical processes. These processes are generally characterized by early technological maturity, and regulatory uncertainty appears to be the most critical issue in limiting the investments (IRENA, 2019). However, different countries are setting specific targets for advanced biofuels in the next years, and thus an increase in production is expected.

2.3.2   Alternative Biofuels—Renewable Natural Gas

While liquid biofuels are currently representing most of the bioenergy in transport, some countries are evaluating the possibility of exploiting renewable methane, obtained through the upgrading of biogas produced by anaerobic digestion of agricultural or landfill wastes. Biogas plants are already a mature technology, applied in different countries to generate electricity and heat. Since biogas is generally a mix of methane and carbon dioxide, the upgrading to renewable methane would involve a process to separate the carbon dioxide and possibly store it or use it in other applications. One of the advantages of renewable methane is the possibility of exploiting existing natural gas infrastructure for transport, as well as natural gas vehicles that are already in operation in different countries worldwide, including China, Iran, India, Pakistan, Argentina, and Italy.

Biomethane is seeing a continuous increase in Europe, where its generation reached 19.4 TWh in 2017, up from 2.3 TWh five years earlier (EBA, 2019). The number of plants reached 540 in 2017, although with a lower increase, due to the operation of larger plants in the last years. The countries with most plants are Germany, Sweden, the UK, and France, although considering the number of plants per inhabitant Sweden, Iceland, Denmark, and Switzerland take the lead. The potential remains very large, since there are currently almost 18,000 biogas plants in Europe, mostly used for power generation. A shift toward biomethane plants is generally seen as an opportunity to substitute the significant use of imported natural gas in different countries, although in some cases (e.g., Italy) dedicated policies are supporting the use of biomethane in the transport sector.

On the other hand, like every use of methane, attention is needed to avoid any gas leakage, which would have severe consequences on climate change issues. The emission of 1 kg of methane is equivalent to 28 kg of CO_2 under the common assumption of considering 100 years of time frame, but this figure rises to 84 kg of CO_2 if we consider only 20 years, meaning that CH_4 emissions will have a more significant impact in the short term (due to their average lifetime in the atmosphere).

2.4 Emissions of Available Technologies

In the previous section, we highlighted the main advantages and limitations of each decarbonization solution that needs to be evaluated by considering several aspects. On the other hand, to assess the potential benefits of each solution, it is important to compare the specific GHG emissions of the different sources, considering all the phases of the supply chain. Figure 2.3 shows the specific CO_{2-eq} emissions of different technologies for cars, based on a Life Cycle Assessment (LCA) approach. The plot is based on a literature review of multiple research works published recently, and each point represents a specific value for each technology. The limits of the boxes represent the first and third quartile of each distribution (i.e., they are representing the 25% and the 75% of the values), while the line in between represents the median value. The colors of the points represent the reference year that has been used for the study, which has an impact of the energy mix and the performance of the available technologies.

The strong variability of the points is due to the multiple aspects that are associated to the manufacturing and operation of the vehicles, as well as to the supply chain of each energy carrier. The variability is particularly high for electricity- and hydrogen-powered cars, due to the different sources that can be used for the energy generation. While they have both the potential to reach very low emission levels, the current electricity mixes and the hydrogen production from fossil fuels are still limiting the benefits that can be reached. For this reason, it is important to highlight that the deployment of such technologies needs to be coupled to an increase of low-carbon options for both electricity and hydrogen production.

Similar patterns are expected for other vehicles (buses, trucks, etc.), although the research works are mostly focused on cars, due to their higher impact in most countries and to the availability of a wider range of technological options. High-duty

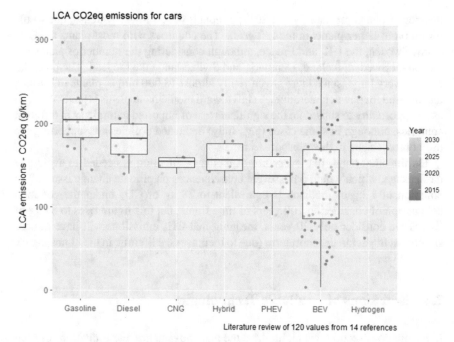

Fig. 2.3 Comparison of specific LCA emissions for cars based on different energy sources (CNG: compressed natural gas, PHEV: plug-in electric vehicle, BEV: battery electric vehicle)

vehicles are generally less suitable to use battery, especially for long-haul operations, due to their weight and space. The use of hydrogen may be a better option, although its efficiency on the entire value chain may remain lower than the direct use of electricity.

2.5   Other Decarbonization Measures

In parallel to alternative powertrain technologies, it is important to remind that other actions are needed toward a better use of vehicles, by optimizing their operation and increasing the average number of passengers. Private cars are often used by a driver without any passenger, especially when commuting for work. The implementation of carpooling services or an increased use of public transport may lead to strong emission savings without the need of investing in new technologies and supply chain. Effective sustainable mobility planning could play a strong role in transport decarbonization, just like energy efficiency is crucial in supporting the current transition toward a low-carbon energy system.

In the short-term, the adoption of more efficient vehicles can play a strong role in decreasing energy consumption in different transport segments. The continuous technological development leads to significant improvements for new vehicles both for fuel consumption (and thus carbon emissions) as well as for pollutants abatement systems. This happens for cars, trucks, buses, and planes. Regulations and standards are continuously pushing manufacturers to improve their products, especially in developed countries. Unfortunately, the lack of regulations in most African and Asian countries is creating a large re-use of old vehicles that are no longer allowed in their country of origin. While re-use may save primary resources for manufacturing new vehicles, the additional fuel consumption and local environmental impacts may offset this advantage.

Significant savings for energy consumption and carbon emissions can be obtained through modal shifts from private cars, especially in urban environments. In particular, first- and last-mile trips can happen by walking of cycling, provided that separate bike lanes and walking paths are available to ensure the safety of the citizens. Active modes can be coupled with public transport to increase the efficiency of the mobility system, especially where electricity-powered solutions are available (e.g., light rail or subway systems).

Moreover, cars can be used more efficiently by increasing their average occupancy, which currently stands around 1.5–1.6 passengers per vehicle (including the driver) but drops to 1.1–1.2 for work trips. Carpooling schemes may help increasing these figures, when properly supported by sustainable mobility polices. Digital technologies may unlock successful business models by helping matching demand and supply of carpooling through dedicated web platforms (see Chap. 3).

2.6   Conclusions and Key Take-Aways

Transport remains among the most carbon-intensive sectors, due to its strong dependency on oil products, currently representing around 92% of final energy consumption in transport worldwide. Moreover, mobility demand is expected to increase significantly in the next decades, especially in Asia and Africa, leading to a huge rise of energy consumption and carbon emissions if no action is performed. Different technological options are available to help decarbonize the sector, and while many of them have already demonstrated a technical maturity, their cost is still higher than conventional fossil-based alternatives.

Electric vehicles are being seen as the most promising solution by many countries and car manufacturers, and thanks to specific policies EVs sales are showing promising trends in China, the USA, and Europe. Western countries are mostly focused on passenger cars, while in China electrification is affecting also buses and two-wheelers. However, to unlock the full decarbonization potential of EVs, proper measures are needed to increase the share of low-carbon sources in the electricity generation mix. While electricity may be a convenient source for light vehicles that need limited ranges, it may not be suitable other sectors such as trucks, long-haul flights, and international shipping.

A potential complement to electrification may be the use of fuel cell vehicles powered by hydrogen, which can guarantee longer ranges and shorter refueling times. Just like electricity, hydrogen needs to be produced in a sustainable way from low-carbon sources, such as electrolysis powered by low-carbon electricity (green hydrogen) or natural gas reforming coupled to carbon capture use and storage systems (blue hydrogen). Current costs for hydrogen are not yet competitive, both for vehicles and supply chain, but manufacturing upscale may bring them down at the level of other transport solutions.

A third solution that is already commercialized since decades is represented by biofuels, although they are currently blended with traditional oil products. Biofuels, when produced in a sustainable way, will play a crucial role in the decarbonization of particular transport segments, especially aviation and shipping. Some challenges remain for the development of advanced biofuel pathways, which are not based on crops in competition with food or feed applications.

Finally, while there are different solutions being evaluated for the supply side, proper actions need to be taken on the demand side to use the available vehicle fleets in the most efficient way possible. Public transport and active modes should be preferred when available, especially in urban environments, and shared mobility options should be supported to increase the average vehicle occupancy and utilization rate.

While decarbonization policies will be crucial for tackling climate change, other aspects are involved in sustainable transport, including local pollution, congestions, noise, equitable access to mobility options, land use and safety of citizens. For this reason, it is important that policies are coordinated and discussed at multiple governance levels, as well as correctly communicated to citizens, to maximize their effectiveness and support a more efficient and sustainable transport system.

References

AFC TCP. (2019). *AFC TCP 2019 survey on the number of fuel cell vehicles, hydrogen refueling stations and targets*. Retrieved from https://www.ieafuelcell.com/fileadmin/publications/2019-04_AFC_TCP_survey_status_FCEV_2018.pdf.

Bachman, J., Fan, R., & Cannon, C. (2018, January). Watch out, airlines. High speed rail now rivals flying on key routes. *Bloomberg*. Retrieved from https://www.bloomberg.com/news/articles/2018-01-09/high-speed-rail-now-rivals-flying-on-key-global-routes.

Barnstone, D. A., & Barnstone, R. V. (2018). The electric highway: Intelligent infrastructures for kinetic cities. *Urban Energy Transition*, 153–166. https://doi.org/10.1016/B978-0-08-102074-6.00022-X.

ClimateWire. (2009, July 1). Sinking finances throw Iceland's "Hydrogen-Based Economy" into the freezer. *NY Times*. Retrieved from https://archive.nytimes.com/www.nytimes.com/cwire/2009/07/01/01climatewire-sinking-finances-throw-icelands-hydrogen-bas-47371.html?pagewanted=all.

COAG Energy Council. (2019). *Australia's National Hydrogen Strategy*. Australia. Retrieved from https://www.industry.gov.au/data-and-publications/australias-national-hydrogen-strategy.

EBA. (2019). *EBA statistical report 2018*. Retrieved from https://www.europeanbiogas.eu/wp-content/uploads/2019/11/EBA_report2018_abriged_A4_vers12_220519_RZweb.pdf.

Hanley, S. (2019, June). Good news about electric trucks coming from Oregon & Switzerland. *Clean Technica*. Retrieved from https://cleantechnica.com/2019/06/24/good-news-about-electric-trucks-coming-from-oregon-switzerland/.

Heid, B., Linder, M., Orthofer, A., & Wilthaner, M. (2017). Hydrogen: The next wave for electric vehicles? *McKinsey*. Retrieved from https://www.mckinsey.com/industries/automotive-and-assembly/our-insights/hydrogen-the-next-wave-for-electric-vehicles.

Hydrogen Council. (2017). *Hydrogen scaling up*. Retrieved from https://hydrogencouncil.com/wp-content/uploads/2017/11/Hydrogen-Scaling-up_Hydrogen-Council_2017.compressed.pdf.

IEA. (2019a). *CO_2 emissions statistics*. Retrieved August 8, 2019, from https://www.iea.org/statistics/co2emissions/.

IEA. (2019b). *Global EV Outlook 2019*. Retrieved from www.iea.org/publications/reports/globalevoutlook2019/.

IEA. (2019c). *IEA CO_2 emissions from fuel combustion statistics*. https://doi.org/10.1787/co2-data-en.

IEA. (2019d). *Renewables 2019*.

IEA. (2019e). *The future of rail*. Retrieved from https://webstore.iea.org/the-future-of-rail.

IEA. (2019f). *World Energy Outlook 2019*. https://doi.org/DOE/EIA-0383(2012) U.S.

IRENA. (2019). *Advanced biofuels—What holds them back?* Retrieved from https://www.irena.org/-/media/Files/IRENA/Agency/Publication/2019/Nov/IRENA_Advanced-biofuels_2019.pdf.

Koltsaklis, N. E., Dagoumas, A. S., & Panapakidis, I. P. (2017). Impact of the penetration of renewables on flexibility needs. *Energy Policy, 109*, 360–369. https://doi.org/10.1016/j.enpol.2017.07.026.

Levy, A. (2019, January). The verdict's still out on battery-electric buses. *Citylab*. Retrieved from https://www.citylab.com/transportation/2019/01/electric-bus-battery-recharge-new-flyer-byd-proterra-beb/577954/.

Liu, K., Wang, J., Yamamoto, T., & Morikawa, T. (2018). Exploring the interactive effects of ambient temperature and vehicle auxiliary loads on electric vehicle energy consumption. *Applied Energy, 227*, 324–331. https://doi.org/10.1016/J.APENERGY.2017.08.074.

Liu, Z., Kendall, K., & Yan, X. (2019). China progress on renewable energy vehicles: Fuel cells, hydrogen and battery hybrid vehicles. *Energies, 12*(1), 1–10. https://doi.org/10.3390/en12010054.

Nagashima, M. (2018). *Japan's hydrogen strategy and its economic and geopolitical implications* (Etudes de l'Ifri). Retrieved from https://www.ifri.org/en/publications/etudes-de-lifri/japans-hydrogen-strategy-and-its-economic-and-geopolitical-implications.

Sinsel, S. R., Riemke, R. L., & Hoffmann, V. H. (2020). Challenges and solution technologies for the integration of variable renewable energy sources—A review. *Renewable Energy, 145*, 2271–2285. https://doi.org/10.1016/J.RENENE.2019.06.147.

Slowik, P., & Lutsey, N. (2019). *The surge of electric vehicles in United States cities* (icct Brief). Retrieved from https://theicct.org/publications/surge-EVs-US-cities-2019.

Statistics Japan. (2019). *Automobiles registered*. Retrieved December 28, 2019, from https://stats-japan.com/t/kiji/10786.

Statistics Norway. (2019). *Electricity statistics in Norway*. Retrieved December 28, 2019, from https://www.ssb.no/en/elektrisitet.

Chapter 3
Digitalization Trends

Abstract In parallel with the policy-driven decarbonization trend discussed in Chap. 2, the transport sector is currently affected by a bottom-up strong technological push by digital technologies at different maturity levels. Potential game changers include Mobility-as-a-Service, shared mobility, autonomous vehicles, and other effects of extra-sector digital technologies (e.g., online platforms, virtualization, e-commerce). This chapter will be dedicated to discussing the possible impact of these technologies in the transport sector, with a focus on the urban context, which is the first being affected by these changes.

3.1 Introduction

Digital technologies are proving to be a game changer in multiple domains, thanks to the possibility of supporting new business models and increasing the efficiency of traditional schemes. The availability of huge computational power and vast amount of data gathered from different sources helps the development of accurate algorithms and the deployment of new services for the customers. Today, a large part of the population worldwide is connected to the internet, and an increasing share of traffic is related to mobile connections. People use smartphones to connect to friends, watch movies and videos, listening to musing, shopping online, finding travel, and accommodation solutions. Smartphones are gradually becoming a key tool to access a wide range of services, and the possibilities offered by digital technologies are in continuous evolution.

Transport is no exception. Web platform is being used, just as in other sectors, to support different sharing economy solutions. Urban citizens in different parts of the world can access shared fleets of cars, bikes, and scooters through their smartphone, or they can just opt for a driver to pick them up to their desired destination. The users may be even able to compare different mobility options, to find the most cheap, fast or comfortable solution for a given trip in any given moment, based on live data on the traffic conditions. Mobility is increasingly being seen as a service, and the traditional model centered on private cars is being challenged by the possibility of paying for the actual mobility needs rather than owning a personal vehicle, with

© The Author(s) 2020

M. Noussan et al., *The Future of Transport Between Digitalization and Decarbonization*, SpringerBriefs in Energy, https://doi.org/10.1007/978-3-030-37966-7_3

potential advantages for cost and convenience. Furthermore, the future deployment of connected and shared automated vehicles may completely change the way people move.

Technological development is not the only aspect. There is a parallel change of mindset, especially for younger generations, which may have different habits, values and needs. The users' behavior is a key for the success or the failure of different mobility models (Cohen, 2019), and there is a large segmentation of preferences and choices across countries, gender, age, and income. National and local policies will also have a crucial role in supporting specific technological solutions, mainly driven by the potential advantages that they could provide to citizens.

However, an additional aspect to be considered is the potential digital divide triggered by these innovative solutions, which can exacerbate the difference among low-income and high-income classes, young and old generations, urban and rural citizens. These growing dualities may be caused by differences in access to technologies related both to financial availability, but also to the digital know-how of different classes of citizens (including literacy, gender, and age).

This chapter will consider the main ongoing trends supported by digital technologies, including Mobility-as-a-Service (MaaS), shared mobility, autonomous vehicles, data-driven mobility planning, and external digital trends that have an indirect effect on mobility demand and behaviors.

3.2 Mobility-as-a-Service—A New Way of Thinking?

Mobility-as-a-Service (MaaS) is based on the possibility of exploiting digital platforms to support the users in choosing multimodal trips, to get the most from the opportunities provided by each mode to fulfill specific requirements. A key aspect is to provide to the travelers updated information on the alternative possibilities, increasing the flexibility of the available options and the resilience of the trip planning against potentials delays or the need of changing itinerary during the trip itself. To provide such information a complex integrated digital platform is required, to collect all the available data from different sources and provide the users with a clear and synthetic dashboard containing all the relevant information to support their trip planning and re-planning when necessary.

The implementation of an effective MaaS platform, which is starting to be tested in different cities worldwide, requires as backbone an efficient public transport system, on which alternative modes could be integrated, including shared mobility, taxi cabs, and active modes. MaaS can be developed at different levels, from a basic system that help the users to compare the available travel possibilities, to a totally integrated environment where the customers can buy each travel solution through a single account, or even pay flat rates to have access to an unlimited number of trips over a certain time period (e.g., monthly). While the former level is already available from multiple platforms in different cities worldwide (e.g., Google Maps, Moovit,

Table 3.1 Available pricing plans for Helsinki, November 2019

Plan	Whim Urban 30	Whim Weekend	Whim Unlimited	Whim to Go
Price	€59.7 (30 days)	€249 (30 days)	€499 (month)	Pay as you go
Public transport	30-day ticket	30-day ticket	Unlimited single tickets	Pay as you go
City bike	Unlimited	Unlimited	Unlimited	Not included
Taxi (5 km)	€10	−15%	Unlimited	Pay as you go
Rental car	€49/day	Weekends	Unlimited	Pay as you go

Source MaaS Global (2019b)

Citymapper), the latter is currently being tested in some cities to gain experience and verify the effect of users' behavior.

Probably, the most interesting case study today is Whim, an app developed by MaaS Global and initially operated in Helsinki, Finland (MaaS Global, 2019b). The app allows the users to plan their journeys over different modes (including public transport, taxis, bike, and car sharing) through their single platform, and propose different pricing schemes, from pay-as-you-go to all-inclusive plans based on a monthly fee. Whim is in operation in Helsinki from the end of 2017, and is currently in available also in Birmingham, Antwerp, and Vienna, although still with limited pricing plans. The current pricing plans available in Helsinki are reported in Table 3.1, to give an idea of the available options and their price. As a comparison, a public transport single ticket in Helsinki costs €2.80, and a 30-day ticket €59.70 (or €53 if the customer chooses an annual subscription), which is the very same price of the Whim Urban 30 pricing plan. While these plans are still at an early phase, the economic sustainability of MaaS business models will need to be demonstrated in the long run.

The company has recently completed a new funding round, including funds from BP Ventures and Mitsubishi Corporation, with the aim of expanding their market to other European cities, Singapore, Tokyo, and North America in 2020 (MaaS Global, 2019a). Whim reached a total of over 6 million trips from its launch, considering all the cities in which is operating. However, many questions related to the potential effects of MaaS still wait for an answer. The key point is if MaaS will be able to shift users from private cars to more sustainable transport modes, or if indeed if all-inclusive pricing plans will lead to a rising transport demand for less effective solutions.

A white paper on the analysis of operational data of 2018 in Helsinki gives some preliminary insights (Ramboll, 2019), although the results are based on the first year of operation, which includes a continuous increase of both registered users and available modes. Some results show that the total number of trips remains the same (an average of 3.3 daily trips), but the largest modal shift in Whim users is from active modes (walk and bicycle) to public transport. This effect may increase the quality of life of the users, by allowing them to save time of travel more comfortably, but at the same time it is increasing the transport demand for motorized modes. Car usage seem to remain the same across the two user groups, but to clarify this aspect additional research is needed on a larger set of data, preferably across different cities.

While MaaS is currently an option for early adopters, its extension to a wider global audience will probably need to face additional issues, including the potential digital divide across generations, the interaction with local regulations in different countries, the ownership of the platforms (public vs. private), potential issues with privacy for users data (as will be discussed in more detail in Sect. 3.5). As already discussed, the potential success of MaaS will be linked to the possibility of demonstrating its capability of triggering a shift toward sustainable transport.

A crucial aspect for the future deployment of MaaS in different cities will be the choice of the business model, and in particular how the digital platform will be built, and which stakeholder will be in charge of operating it. Three basic models are available, with different actors as integrators and multiple aspects to be considered when choosing a solution over another, including alignment with policy goals, market penetration, social inclusion, innovation, customer orientation, impartiality, and data availability for public authorities (UITP, 2019). These models are summarized in Fig. 3.1. The first option is the implementation of a free market open to different MaaS operators, which can define individual agreements with transport operators. This solution may lead to a high degree of innovation and customer service, but with high perceived risks for social inclusion and impartiality, as well as limited data transparency toward public authorities to support local policies. On the other

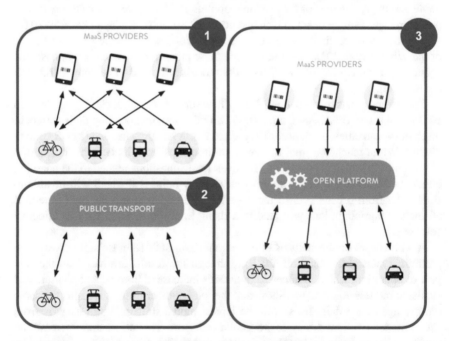

Fig. 3.1 Different MaaS models depending on the type of aggregator. Authors' elaboration from UITP (2019)

hand, the second solution is based on the idea of exploiting the already available public transport operator to directly act as mobility aggregator, by integrating other modes. While this would obviously lead to higher guarantees in terms of social rights, impartiality, and policy support, the innovation may be significantly slowed down, as well as the potential competitivity and attention to customers. A third model, which is somewhat a hybridization of the previous two, would be based on a common open platform defined by clear communication standards, as a backend for different MaaS providers competing on the frontend services to the users. However, while this model provides an interesting opportunity to combine the advantages of the other two, the financing required to develop and operate the open platform may be a significant issue (UITP, 2019).

A final aspect related to MaaS, which is also a result from the first operational data, is the importance of being supported by a reliable and high-quality public transport service, which should represent the backbone of urban mobility, coupled to other transport modes for first-mile and last-mile solutions.

3.3 Shared Mobility—Sharing Assets or Trips?

The concept of shared mobility embraces a large variety of technologies and mobility models, from car sharing, bike sharing, and other shared vehicles (e.g., electric scooters) to ride-hailing and carpooling, which in turn involves the sharing of the very same trip by multiple users. These models have a different diffusion in world regions, depending also on existing contexts and specific problems (such as population density, income levels, pollution, and congestion levels etc.). However, they are generally appearing in urban contexts, especially in large cities where the high density of inhabitants allows a more interesting economic profitability.

The most significant aspect that diversifies shared mobility options is whether they are based on the sharing of a vehicle at different times, or rather on the aggregation of different users that need to do the same trip. All the shared mobility options have in common the aim of going beyond the usual mobility model centered on private car ownership, either by providing the users with alternative transport modes, or by increasing the average load factor of cars, that usually remains well below two people per car in many developed countries.

3.3.1 Car Sharing

Car sharing has seen a large number of applications in the second half of the twentieth century, both in Europe and North America, with the roots of car sharing model dating back to 1948, in Zurich (Shaheen, Sperling, & Wagner, 1998). These mobility models, as alternatives to car ownership, were developed to increase the usage of cars and make them more profitable. Moreover, most companies were publicly backed, with the aim of fostering societal benefits such as lower parking needs as well as expected lower car usage related to the different pricing mechanism. In fact, while car ownership is based on high fixed costs and relatively low operational costs (especially in the USA, where fuel costs are usually lower), the car sharing pricing mechanisms were based on an annual fixed fee and a variable fee related to the actual car usage, with the aim of discouraging an excessive use of the cars. Car sharing models were generally attractive to users with a medium annual car mileage, since occasional users were more attracted from car rental (discouraged by car sharing's annual fee), and for users with high usage, a private car was more convenient. However, car sharing involved additional benefits, such as no ownership responsibilities and the potential access to different cars sizes based on the specific purpose of the trip (when available in the car fleet).

Car sharing models evolved throughout the decades thanks to different technological innovations, especially in matching the available car fleets with the users' demand. This aspect was significantly improved with digital technologies in the last decade, thanks to the possibility of easily checking in realtime available cars based on the position of the user, and to go straight through the booking and paying process. This technological development, in parallel with local policies supporting lower private cars usage due to environmental concerns, has unlocked new business models, pushing different companies to offer free-floating car sharing fleets to travelers, especially in large cities.

Car sharing operators claim different advantages in comparison with privately owned cars, including the higher utilization rate, the lower emissions thanks to a faster fleet replacement, lower parking needs thanks to higher user-per-car rates. However, a key aspect is whether car sharing is able to trigger a shift from private cars or if it ends up in providing the possibility of using a car to people that were used to rely on public transport. This latter case may reflect an increased quality of life for citizens, but with higher environmental impacts. Research works have found mixed results, depending on the specific case study. A key driver appears to be the interaction with public transport: cities with car sharing services oriented to provide first- and last-mile solutions in integration with available public transport are more likely to gather environmental benefits. On the other hand, car sharing often represents an interesting alternative to public transport, providing better flexibility and comfort, although at a higher price. Local policy actions are crucial to foster an integration with public transport, by defining specific regulations for the operation of car sharing and also by implementing MaaS platforms to provide the users with a single platform for multimodal trips (see Sect. 3.2).

3.3.2   Ridesharing

Ridesharing companies, also known as ride-hailing services, provide mobility services that are similar to taxicabs, by matching passengers with drivers exploiting websites or mobile applications. They have similar business models of other sharing economy companies, limiting their activity to the ownership and management of the digital platform, without owning any vehicle or other significant assets. For this reason, ridesharing companies have been able to operate without the needing of respecting the taxicabs regulations in many countries, resulting in lower operation costs leading to lower prices for the final users. Moreover, they were also able to provide access to mobility to poor or isolated neighborhoods that are not regularly served by taxicabs. However, many countries and cities have implemented specific regulation to level the playing field among ride-hailing and traditional taxicabs, and in some jurisdictions, they have been even banned.

Web-based ridesharing companies appeared in the last decade in the USA, and the two most notable examples are now Uber and Lyft, which have been in a continuous competition since their creation in the early 2010s. Since both consider drivers as independent contractors rather than employees, in many cities several drivers are registered on both platforms. Uber and Lyft are both characterized by aggressive business models, which resulted in several criticisms on different aspects, mostly complaints from taxi drivers for illegal competition, and lawsuit filed by drivers that were not receiving adequate compensation. These problems have pushed several jurisdictions to develop specific regulations for ridesharing operation, mainly at the city level.

The evolution of Uber and Lyft is an example of how digital technology can support disruptive solutions in transport, although the benefits are still to be proven. Both companies are still showing significant losses (0.9 b\$ for Lyft and 1.8 b\$ for Uber in 2018), and a change of this trend would require either higher fares for passengers or (even) lower compensation for drivers, which are already often earning revenues lower than minimum wages after taxation. Moreover, their users are mostly shifting from public transport rather than private cars, with negative effects on urban traffic and pollution. Both companies have introduced the possibility of sharing rides, to switch toward a more sustainable use of their vehicles. However, research studies in New York and San Francisco have demonstrated that this effect remains limited, and the net effect of ridesharing services is an increase of urban traffic. In particular, one-passenger ridesharing services add up 2.8 vehicle km on the street per each vehicle km of personal driving removed, and the inclusion of shared services marginally decreased this figure to 2.6 vehicle km added (Schaller Consulting, 2018). The main reason is that most users switch from non-auto modes, including public transport, walking, and cycling. In addition, there is added mileage between trips as drivers wait for the next dispatch and then drive to a pick-up location, and even in a shared ride, a part of the trip involves just one passenger when pick-up locations are different.

However, as it happens with many other trends, it is important to keep in mind that other world regions have very different contexts, and the considerations for Europe or North America may not be applicable. In particular, in large cities without well-developed public transport solutions (e.g., Asia, Africa, South America) ridesharing can provide an economically sound solution for citizens that cannot afford a private vehicle but has the resources to pay for a higher-level service compared to three-wheelers or mopeds. Moreover, other aspects are important in the analysis of the ride-hailing business model, such as the ownership of the vehicles. While in the USA, ride-hailing services are often provided by private citizens with their personally owned cars, as a part-time activity, this is not the case for other countries, such as India. The cars registered on Uber or Ola (another ride-hailing app) platforms are often owned by large transport contractors that hire drivers, whose earnings are as low as one tenth of those of drivers owning their personal vehicles (Karnik, 2017). However, the investment costs of a new car remain a strong barrier for most drivers, although ride-hailing companies are pairing up with banks to provided dedicated loans. These aspects are of crucial importance for the policy makers that need to assess all the social and economic benefits supplied by ride-hailing platforms to citizens.

3.3.3 Carpooling

A better approach for sustainable transport is the concept of carpooling, which aims at increasing the average number of passengers per car by matching users that have the same travel demand. Some carpooling schemes are dedicated to daily commuting to work, which shows a significant potential due to the regular frequency of the trips and to the average car occupancy, which is much lower than for leisure trips, with most car owners driving alone to work. Carpooling has been widely adopted to share the operating costs of travelling by car, with significant benefits related to fossil energy savings, lower CO_2 and pollutants emissions, less traffic on the roads and lower needs for parking spaces.

Whereas carpooling schemes have often been informally organized, mainly between family members or colleagues, the availability of digital platforms to match demand and supply has represented a significant opportunity to increase their diffusion. Digital-driven carpooling has also seen a renewed interest for longer one-off journeys, for which it was harder to find potential matching passengers. In long journeys, carpoolers often join for a part of the trip, giving more attractiveness and flexibility to carpooling and allowing and easier matching between drivers and passengers. Online platforms for carpooling usually use community-based trust mechanisms, such as drivers' and passengers' reviews, to increase user comfort and security during the trips.

While all these advantages seem to promote carpooling as a sustainable solution, it should be noted that, like for other transport solution, a key aspect in evaluating its effectiveness lays in the modal shift that is involved in the process. Carpooling is a significant improvement for users that would have travelled with private cars, but when it draws to cars previous public transport users, the benefits are harder to quantify. A recent study has considered the annual operational data of BlaBlaCar, one of the largest web-based carpooling communities, to evaluate its net contribution to CO_2 emissions in selected countries (Butt d'Espous & Wagner, 2019). The results show a positive effect driven by the large increase of average car occupancy rate, which increase from 1.9 to 3.9 in the eight selected countries (it has to be noted that BlaBlaCar provides only trips between different cities, with a very low share of commuting users). The study also considers the actual modal shift of the carpooling users, thanks to dedicated surveys, and it includes the additional trips that would have not been done if carpooling was not available (representing 5.2% of the members, ranging from 2.8 to 9.6% on a country basis). The study is based on 95.3 million journeys in 2018, and the calculated emissions of carpooling result in 2.2 million tons of CO_2, compared to 3.1 Mt that would have been emitted if carpooling were not available.

While this case represents a very positive example of the possible benefits of shared mobility, the actual modal shift and the additional trips generated remain two key aspects that may undermine the sustainability of other carpooling models. Again, proper policies are needed to help developing an integrated approach to mobility planning and avoid direct competition between public transport and private-based mobility models.

3.3.4 Bike Sharing

Bike sharing schemes had a long history, dating back to 1965 in Amsterdam, when citizens organized a system to share free bikes painted in white and made available to users throughout the city. Unfortunately, this program suffered from the lack of dedicated locks that finally made most of the bikes to be damaged or stolen, leading the program to be shut down. Public-based bike sharing schemes took decades to develop, and at the beginning of the century, there were still less than ten systems worldwide. However, the availability of web platforms and the possibility of locating available bikes in real time based on the user location through a smartphone have supported a strong rise in bike sharing schemes in multiple cities worldwide. Furthermore, Internet of things has unlocked the possibility of dockless systems (aka free-floating): since each bike is always connected to the web, thanks to an algorithm it can be locked and unlocked when needed through an app, and it can be left anywhere (or almost) since its position is continuously recorded.

Station-based and free-floating systems represent two distinct models, which have mostly remained separate until now. The former has been generally the result of a choice of each city, with a part of financing provided by public municipalities to foster clean transportation for the citizens. Some systems have been implemented before the digital transition and the diffusion of smartphones, resulting in a bottom-up development of bike sharing systems with the users needing to register to each city separately, often with the need of a physical registration process. Conversely, dockless systems have been deployed across different cities by large companies (mostly Chinese) with a top-down approach, often without a real dialogue with the involved municipalities, and an advantage for the users is that they can access the service in different cities by registering once to the platform. In last years, some hybrid systems are being implemented to try to maximize the advantages for the users.

The bike sharing market worldwide shows a strong expansion, and it is not easy to access updated statistics due to the absence of a common association or data collection standard. The largest amount of bike sharing trips happens in China, where the immense market increase from 2015 has been mainly driven by private players (Roland Berger, 2018) and few public data are available. As of December 2019, the most comprehensive map of existing bike sharing programs across the world (DeMaio, 2019) lists 2120 cities in operation, 368 cities in planning or under construction, and 414 cities that are no longer operating. Another platform (O'Brien, 2019) is performing a real-time tracking of 474 cities worldwide, by exploiting the available data from multiple sources. The service is currently monitoring more than 44 thousand docking stations for 380 thousand bikes, providing detailed information for each city down to the dock level, with historical trends over the last 24 h. Another similar service (Citybikes, 2019) is providing open data information through a dedicated application program interface (API) with live information for more than 400 cities worldwide.

As in other shared mobility modes, the use of data generated by the users is at the center of a business strategy to optimize the operation of the system as well as to provide a better users experience and additional services to the customers (Roland Berger, 2018). The availability of bike sharing operating data, often made available as open data to the public, is a precious resource for policy makers and mobility planners to evaluate the usage of bike sharing in cities, and to plan future improvements as well as verify their effectiveness. In general, bike sharing usage has a strong variation over the day, in line with the usual mobility patterns in cities related to commuting (see the example of the bike sharing of Turin, in Fig. 3.2), but at the same time the seasonal variation related to weather conditions can play a stronger role than for other modes, although with variations from a country to another.

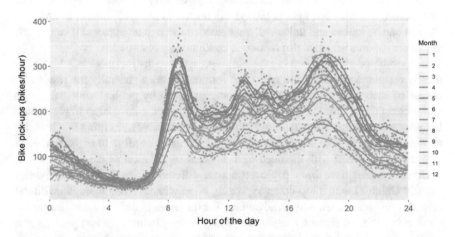

Fig. 3.2 Daily profile of bike sharing use in Turin (based on 5-min data in 2018). *Source* Noussan, Carioni, Sanvito, and Colombo (2019)

Bike sharing has proven to be an effective solution especially for first- and last-mile trips, often coupled to public transport solutions. The growing ecosystem of integrated mobility solutions, driven by MaaS platforms, can further support the evolution of bike sharing, thanks to its potential in supporting the choice of public transport, to discourage the use of private cars when they are not necessary.

In some cases, bike sharing is being equipped with electric bikes, which provide the users with a more convenient solution, especially in cities with hilly terrain. The use of e-bikes generally requires an extra fee for the users, and attention needs to be paid to the charging infrastructures of e-bikes to ensure a proper user experience. The integration of e-bikes in existing station-based bike sharing systems often required the upgrade of the docking stations to allow for the charging of batteries.

3.3.5 Electric-Powered Micromobility

Another phenomenon that is showing a considerable hype in the US as well as in major European cities is the deployment of electric-powered micromobility, notably electric scooters. While bike sharing systems involve the use of a transport mode that has a long history and thus specific rules for its operation over roads, many countries and cities are still facing the need of developing proper regulations for electric scooters, which thus remain in a gray area in many cities and states/countries. Different companies have exploited the lack of regulation to deploy their e-scooters in cities, but in some cases, they have been subsequently forbidden to operate (e.g., San Francisco), due to issues including improper scooter parking on sidewalks and low availability of scooters in low-income neighborhoods.

The story of electric scooters sharing systems starts in the USA, where they have shown a massive deployment in the last few years, with total e-scooters trips in 2018 reaching a value of 38.5 million, higher than the 36.8 million trips of station-based bike sharing systems (NACTO, 2019). As of the end of 2018, the number of e-scooters available in about 100 US cities was over 85,000, but 40% of all e-scooter trips took place in the Los Angeles, San Diego, and Austin regions. These e-scooters have almost totally substituted the use of dockless bike sharing systems, with the notable exception of Seattle, since different bike sharing operators have shifted their focus on e-scooters by retooling their fleets (including Lime and Spin).

An analysis in some US cities on usage patterns and profiles (NACTO, 2019) showed that e-scooters users are more similar to casual bike sharing users that to annual subscribers, the latter using bikes to commute and thus in morning and evening rush hours. Conversely, riders are using e-scooters for different purposes, including social, shopping and other recreational activities. For the same reason, e-scooter ridership shows higher interest during weekends than during weekdays. E-scooters are usually more expensive than bike sharing or e-bikes, with the average trip in 2018 costing to the user 3.50$ to travel little more than 1 mile in roughly 15 min. Thus, many cities have requested the companies to provide discounts to low-income citizens to support a more equitable access to mobility options in the US cities.

The huge use of e-scooters for occasional and very short trips suggests that they have involved a modal shift from other active modes, mostly walking and cycling. Thus, the use of e-scooters involve an additional electricity consumption, together with the need of moving the e-scooters to charging stations at night and then back to the streets at early morning. While the electricity consumed may be produced by renewables, the gasoline burned by the vans used to move the scooters overnight is not. For this reason, the use of e-scooters may not be classified as a sustainable mobility option, unless a better integration with other modes and alternative charging strategies help demonstrate real benefits. On the other hand, it is always important to remember that while sustainability is an important aspect, the supply of additional services to the citizens (including faster and more convenient mobility options) should be in some way accounted for in a systemic perspective that goes beyond the simple evaluation of the energy consumption.

Besides the USA, e-scooters are being deployed in several European cities by different private companies, and the available market reports expect a very strong potential in Asian countries in the next decade, which may represent the largest share of e-scooters sales by 2030.

3.4 Autonomous Vehicles—Would You Bet on It?

Among the different innovations that are currently fostered by digital technologies, self-driving vehicles represent at the same time the less mature and the one with the highest disruption potential. The passenger transport of the last century has been mainly centered on private cars, thanks to the high flexibility and convenience of driving. At the same time, everyday mobility is also characterized by frequent congestions, especially in large urban areas, leading to waste of time, environmental impacts, and safety issues. Autonomous cars may strongly enhance the travel experience of passengers, by providing an even higher degree of flexibility without the need of staying behind the wheel. In fact, passengers may dedicate to a large variety of activities during travels, and the time that is currently wasted can be exploited for working or entertainment. Moreover, the time required to find a parking spot in the city center will no longer be needed, and fleets of shared AVs may continuously remain in movement, based on optimization algorithms that coordinate their paths based on the users' demand.

The key aspect supporting the deployment of AVs is the enhanced safety of road transport, since a total shift toward self-driving cars may remove all the accidents caused by human errors. AVs can also guarantee a smoother operation, with more efficient driving cycles and a better coordination with other vehicles leading to fewer congestions. Still, the interaction between computer-driven cars and human-driven cars (or bicycles, pedestrians, animals) may be less easy to manage, since humans are much less predictable, and the absence of eye contact may lead to difficult interactions between humans and robots. Thus, the transition phase toward the use of AVs may prove to be even more complicated, and potential issues may also discourage people to move away from their current traditional cars.

Autonomous cars are usually classified based on the level of automation that they can provide, on a scale from level 0 (no automation) to level 5 (full automation). These levels represent a gradual increase of the tasks that are carried out without the manual intervention of the driver, including accelerating/decelerating and steering, but also the range of conditions in which the car can handle an autonomous operation, with particular attention to weather issues, such as rain and snow. While a full automation needs to handle a wide range of aspects, a partial automation has the additional issue of managing the interaction with the driver, and the switch of controls between car and driver. In particular, the driver may be asked to intervene in emergency situations, but his reactivity may be strongly penalized if he is not already paying attention to what is happening around them.

Research and development of AVs require huge investments, and automotive companies are often developing joint projects with other competitors to share the efforts and the potential risks. Many IT companies are also developing their own solutions, by exploiting their know-how to define proper algorithms and control systems. Two opposite strategies are possible to reach the full automation, either by heading directly toward level-5 automation, or by adding gradual levels of automation. The former strategy involves higher technical barriers and may require waiting a long time before producing a commercial car, while the latter allows to monetize each additional improvement by continuously upgrading the commercial vehicles on the market. However, this slows down the entire process, and focusing too much on single improvements may leads to never reaching the final goal.

The operation of AVs requires a huge amount of data collection and elaboration, both from onboard sensors and from a continuous communication with other vehicles (vehicle-to-vehicle, V2V) or with the external infrastructure (V2I). Thus, a significant improvement of mobile data infrastructure will be required, since the deployment of connected cars will lead to a quick exponential increase of data flow. Each data communication will also need to face issues related to privacy concerns and potential cybersecurity threats. Moreover, the onboard hardware and software solutions will need to be able to handle increasing amounts of data, which will increase by orders of magnitude. Today a connected car, even at a low degree of autonomy, is generating roughly 25 GB per hour. These numbers can rise to between 1.4 and 19 TB per hour for an autonomous car, depending on the sensors setup (Dmitriev, 2019).

Apart from the technical aspects, the development of AVs will also involve ethical issues. Since the AVs operation will be based on human programmed algorithms, in the eventuality of an unavoidable collision with pedestrians, the car needs to be programmed to choose whether to protect the passengers onboard or the pedestrians outside. While human drivers often react in an unpredictable way in emergency situations, AVs will just apply precisely the set of rules that they have been taught. It will not be trivial to reach a consensus on these ethical issues, especially considering the huge variety of situations that may occur. Moreover, different cultures may have different approaches and opinions on what is ethical and what is not. Finally, in the event of an accident involving AVs, it is not clear who will take responsibility for it:

national laws as well as insurance companies' regulations will need to be modified to handle these potential situations.

Given all these issues and opportunities, a large uncertainty remains on the eventual effect of AVs on the energy consumption and the environmental impacts of the mobility system. The automation of vehicles may improve driving cycles with strong benefits for fuel economy, and the potential reduced congestions will also increase the system efficiency. However, these may be outpaced by a strong increase in demand driven by the high quality of travelling in self-driving cars, even by people that are currently using public transportation. Moreover, a group of people that are not able to drive (children, elderly or impaired) may grant access to mobility and create additional demand. The higher productivity during travel may also decrease the issues related to commuting time, pushing the people toward an even increased urban sprawl to benefit from lower housing prices far from city center.

One of the key aspects for energy consumption will be the ownership model, and in particular, if AVs will be privately owned of shared. A private autonomous car will be more similar to the use of the car we have today: we are able to choose the model we prefer, we can store items in it and in some cases it represents a status symbol. On the other hand, it will end up in being much inefficient, just as today's cars: it will necessarily involve a significant amount of empty trips to look for a parking when not needed, and it will still be useful for a very limited amount of time. Conversely, a fleet of shared AVs may be more effective in optimizing the mobility demand of different people, with the possibility of also sharing trips with other passengers. Empty vehicles may be kept in motion, waiting for other passengers requesting a vehicle. However, while this may decrease the need for parking in city centers, loading and unloading zones will still be needed, and the continuous traffic of empty vehicles nearby popular locations, required to ensure a timely response to people's mobility demand, may end up to be a backlash for both safety and environment. Users' behavior and mobility patterns will remain at the center of effectiveness of shared travels. While AVs may provide optimized solutions for occasional mobility, the systemic travels, and especially commuting, are harder to improve. Today's mobility demand in cities is often characterized by strong peaks both in the morning and in the evening, caused by commuting trips. Without any major changes, AVs will provide little benefits for congestion, unless people accept to share the vehicles with other passengers.

Nowadays, there are many companies involved in testing autonomous vehicles worldwide, and although there are still no commercial vehicles able to provide full autonomy, some companies are testing full-autonomous ride-hailing services (although with safety backup drivers, in some cases also due to regulatory issues). Testing phases in real conditions are necessary to evaluate the response of algorithms in a wide range of cases that is difficult to predict and replicate in lab. Vehicle test is also important for users, which need to experience the actual behavior of riding a self-driving car.

Finally, while much attention is given to autonomous cars, another interesting application is related to autonomous trucks. Its implementation may be easier, especially in dedicated environments, since their operation is much more regular. The automation of road freight transport may provide significant fuel savings through platooning, i.e., the possibility of driving vehicles together with a minimum distance between each other, thus reducing the air resistance and also increasing the trucks density in highways.

3.5 Data-Driven Mobility Planning

An additional aspect to be discussed is the possibility of exploiting the available data generated by different mobility modes to improve the urban planning. Data can be sourced from a wide range of devices, both related to vehicles (e.g., car sharing, micromobility vehicles) or to users (mainly through smartphones applications), and they can be fitted to different applications. Available data from different shared mobility models can help local planners, as well as company managers, to optimize the distribution of available vehicles in the city neighborhoods, to ensure the highest access to mobility through different classes, especially for micro-mobility options. On the other hand, useful data from smartphones can help to build up reliable models for congestions management and analyze the behavior of people through different modal shares, which can be estimated from the average speed and other aspects related to each trip (such as the number of people sharing the same positions over time, the matching with available public transport schedules, etc.).

In some cases, private companies can provide a support to local communities in transport planning, by sharing their data on users' behaviors and mobility patterns. The use of mobile phones data to support transport planning has seen a dramatic rise in the last decade, with the result of improving the accuracy of the transport monitoring and modelling. Moreover, while these data are generally unable to provide detailed information on the transport mode which is being used, data from specific applications may provide additional insights. An example is the use of fitness apps that register the route of runners and bike riders to evaluate the actual demand for mobility. One of those apps, Strava, has already supported local mobility planning in more than 300 communities, including Portland and Seattle, by exploiting the data provided by more than 45 million users to evaluate the effect of different mobility choices, such as estimating the effect of different bike lanes layouts (Andrews, 2019). Data are provided only to organizations that plan, own or maintain infrastructures, and a proper data aggregation is performed to avoid privacy issues.

However, there are also cases where the problem of data ownership arises, especially if those data represent a significant business value for a company against its competitors. This is often the case for shared mobility players, whose operating data may represent a very valuable resource for urban planners. While there are usually few national regulations, cities are starting to implement specific actions to try to take advantages from available data. One notable example is the Mobility Data Specification, or MDS, an open-source digital tool initially developed by the city of Los Angeles, USA, as a standardized way to collect operational data from dockless e-scooters, bicycles, and car share (Open Mobility Foundation, 2019). This set of interfaces has been developed to help the dialogue between shared mobility companies and municipalities, since currently each site was requiring a different standard with a lack of efficiency for operators. However, some firms, including Uber, are raising concerns about the potential privacy issues for dockless vehicles users, since there is currently no clear statement on the circumstances under which data can and cannot be shared with other agencies requesting it (Zipper, 2019). The central point is whether local planners should receive raw operation data on each single trip of the users, or rather be able to access aggregated data only. Although single trips can be anonymized, research studies claim that even with anonyme data almost each individual can be identified with as little as four space-temporal data points (de Montjoye, Hidalgo, Verleysen, & Blondel, 2013). Effective solutions to ensure users privacy will be crucial for the success of data-driven planning, either by choosing the right level of data aggregation or to ensure a restricted access to those data.

While much focus is on passenger transport, also freight transport can exploit the availability of new datasets to optimize logistics and match the goods demand in the most efficient way. Data related to vehicles, cargoes, drivers, companies, and infrastructures can be used to increase operational experience and to enhance customer experience, and new business models can be created by expanding revenue streams from existing products or by creating new data-driven products. There is a growing trend in companies to exploit digital technologies to try to anticipate consumers' needs, based on a wide variety of historical data collected from customers' behavior. A notable example is the concept of "anticipatory shipping," a patent filed by Amazon in 2013, which involves the shipment of a good before a customer even buys it (Bensinger, 2014). This method, based on machine learning algorithms trained on different data trends, may lead to a significant decrease of shipping costs, since standard shipping is usually much cheaper than two days delivery, but also to a significant increase of customers' experience resulting in higher brand loyalty.

3.6   A Final Look at Digitalization Outside Transport—What Will It Change?

As we saw in the previous sections, digital technologies are enabling multiple mobility solutions, and the future potential is even more significant. In addition, there are also many applications that are outside the conventional boundaries of the transport sector, but which may have significant impacts on mobility demand patterns. Think about the possibility of remote working and the virtualization of goods that can be used without the need of a physical copy shipped to our home (books, music, movies, etc.). These technologies may decrease the mobility demand of workers and goods, lowering the total energy consumption of transport. E-commerce and food deliveries may have a more complex effect on transport, since they involve a shift from passenger to freight mobility, by decreasing some shopping and recreational activities but at the expense of an increase of urban deliveries.

However, while some digital technologies decrease the mobility demand and thus the consequent energy consumption, they require a significant amount of energy to operate the web infrastructure that supports their services. Thus, an energy consumption decrease in transport sector may be compensated or even outweighed by an increase of electricity demand for the entire supply chain of movies or music streaming. Watching a one-hour movie on Netflix causes 3.2 kg of CO_2 emissions (The Shift Project, 2019), which is more than driving 25 km with a new car. Online video streaming is showing an exponential increase each year, and in 2018 it was responsible for 306 Mt of CO_2 emissions, a little less than 1% of total emissions. The trend is expected to continue, since more and more people are accessing these services, and the video quality (i.e., amount of data) is continuously increasing to provide better user experience on larger devices.

The effect of e-commerce is less clear, and different research works conclude that there is no trivial answer at the positive or negative effect of buying things online rather than driving to the mall. Some analyses remark that shopping is seldom the sole objective of a journey, and people combine the need for buying goods with other recreational activities. Thus, lower shopping needs would not directly correlate to a corresponding decrease of mobility demand. Conversely, e-commerce is leading to an unprecedent need of last-mile deliveries in urban environments with commercial vehicles, worsening congestion, local pollution, and fuel consumption (and thus emissions). The one-day or one-hour deliveries push for even higher presence of delivery services nearby the users. And e-commerce may actually increase the total amount of consumed goods, rather than just shifting habits from mall to online shopping.

However, there is still room for optimization, and large companies are exploiting the availability of huge amount of data to discover online buying patterns that may help them predict the future behavior of customers. Amazon has filed a patent on the concept of "anticipatory shipping": the idea of starting to ship any good before the customer buy it (Bensinger, 2014), to cut on delivery times and costs, since slow deliveries cost much less than fast ones. This could even go further, sellers may ship to customers goods that they may like, with the possibility of shipping them back. The efficiency may not seem to be high, but the business model may work, depending on the accuracy of the algorithms and on the additional effort required to customers to ship back a good that they may wanted to buy sooner or later.

As for other digital technologies, the potential of demand increase competes with the potential of a larger efficiency supported by optimization algorithms. It is difficult to predict which trend will prevail over the other, and probably, there will be differences across sectors, regions and times. Any analysis on transport impacts may require a system perspective, since digital technologies are shifting consumption beyond the traditional borders of what we considered to be the transport sector. The transportation of people and good is gradually and partially being shifted to the transportation of information through virtual networks, and this phenomenon needs to be factored in when looking for the future of mobility.

3.7 Conclusions and Key Take-Aways

This chapter depicted a brief synthesis of the main aspects related to digital technologies in transport, by discussing the current state of the art and the possible future evolutions. Digitalization has the potential of becoming a game changer in different sectors, from shared mobility to autonomous vehicles.

At the same time, digitalization interacts with other mobility trends, and the future effect on the demand for transport and the modal share of travelers is far from being clear. The adoption of various solutions will have a large variability from a country to another, and their success will be based both on their capacity to deliver convenient and reliable services to customers at a lower cost, and on their sustainability in terms of environmental impact at local and global scale, urban traffic and safety aspects. Policies and regulations will have a crucial role in fostering the deployment of effective solutions based on available technologies.

Digital technologies are already demonstrating their potential in fostering the use of public transport and shared mobility options, by providing the users with live alternatives for their trips and by supporting the planning and operation of urban transport systems thanks to a wide collection of measured data. Collective transport may decrease the number of private cars on roads, especially in urban environments, leading to fewer congestions, lower need for parking space, better air quality in city centers. These aspects will be more and more important as urban population is continuously increasing worldwide, with the need of a complete rethinking of urban planning to accommodate the related mobility demand of citizens.

A pivotal aspect for future transport systems and models will be the evolution of autonomous vehicles. The effectiveness of AVs still needs to be demonstrated, and in addition to technical aspects, the technology needs to face users' confidence and willingness of adoption and ethical issues related to potential accidents. Finally, even if AVs can reach a significant penetration level, it is not clear if their operation will lead to an optimized use of available vehicles or if they will just increase the total mobility demand and thus the congestion. Policies and regulations will prove to be central in driving a sustainable use of AVs.

A final remark is that digital technologies are not necessarily supporting a more sustainable transport system, although their potential is interesting. Like any other technology, they provide additional opportunities for users and for companies, but their deployment needs to be matched by timely and accurate policies that support a sustainable transport planning by defining clear and achievable targets.

References

Andrews, J. (2019, November). Strava's new tool lets smaller cities unlock their transportation data. *Curbed—Vox Media*. Retrieved from https://www.curbed.com/2019/11/6/20951384/stava-fitness-data-metro-bikes-infrastructure.

Bensinger, G. (2014, January 17). Amazon wants to ship your package before you buy it. *The Wall Street Journal*. Retrieved from https://blogs.wsj.com/digits/2014/01/17/amazon-wants-to-ship-your-package-before-you-buy-it.

Butt d'Espous, V., & Wagner, L. (2019). *Zero empty seats*. Retrieved from https://drive.google.com/file/d/1exHoqlVa3NROt8B92Rulv-BtXbcZaebp/view.

Citybikes. (2019). *Bike sharing data for everyone*. Retrieved December 18, 2019, from https://citybik.es/.

Cohen, K. (2019). *Human behavior and new mobility trends in the United States, Europe, and China* (FEEM Working Papers). Retrieved from https://www.feem.it/m/publications_pages/ndl2019-024.pdf.

de Montjoye, Y.-A., Hidalgo, C. A., Verleysen, M., & Blondel, V. D. (2013). Unique in the crowd: The privacy bounds of human mobility. *Scientific Reports, 3*, 1376. Retrieved from https://doi.org/10.1038/srep01376.

DeMaio, P. (2019). *The bike-sharing world map*. Retrieved December 18, 2019, from http://www.bikesharingmap.com/.

Dmitriev, S. (2019). Autonomous cars will generate more than 300 TB of data per year. *Tuxera Blog*. Retrieved from https://www.tuxera.com/blog/autonomous-cars-300-tb-of-data-per-year/.

Karnik, M. (2017, March). Uber in India is fundamentally different from Uber in the West. *Quartz India*. Retrieved from https://qz.com/india/926220/uber-in-india-is-fundamentally-different-from-uber-in-the-west/.

MaaS Global. (2019a). *MaaS global completes €29.5 M funding round*. Retrieved November 8, 2019, from https://whimapp.com/maas-global-completes-e29-5m-funding-round/.

MaaS Global. (2019b). *Whim app*. Retrieved November 8, 2019, from https://whimapp.com/about-us/.

NACTO. (2019). *Shared micromobility in the U.S.: 2018*. Retrieved from https://nacto.org/shared-micromobility-2018/.

Noussan, M., Carioni, G., Sanvito, F. D., & Colombo, E. (2019). Urban mobility demand profiles: Time series for cars and bike-sharing use as a resource for transport and energy modeling. *Data, 4*(3). https://doi.org/10.3390/data4030108.

O'Brien, O. (2019). Bike sharing map—Global map of bikeshare by OOmap. Retrieved December 18, 2019, from https://bikesharemap.com/.

Open Mobility Foundation. (2019). *Mobility data specification*. Retrieved from https://github.com/openmobilityfoundation/mobility-data-specification.

Ramboll. (2019). *WHIMPACT -Insights from the world's first Mobility-as-a-Service (MaaS) system*. Retrieved from https://ramboll.com/-/media/files/rfi/publications/Ramboll_whimpact-2019.pdf.

Roland Berger. (2018). *Bike sharing 5.0*.

Schaller Consulting. (2018). *The new automobility: Lyft, Uber and the future of American cities*. Retrieved from www.schallerconsult.com/rideservices/automobility.pdf/http://www.schallerconsult.com/rideservices/automobility.htm.

Shaheen, S., Sperling, D., & Wagner, C. (1998). Carsharing in Europe and North America: Past, Present and Future. *Transportation Quarterly, 52*(3), 35–52. Retrieved from https://web.archive.org/web/20120320192219/http://www.uctc.net/papers/467.pdf.

The Shift Project. (2019). *Climate crisis: The unsustainable use of online video*. Retrieved from https://theshiftproject.org/en/article/unsustainable-use-online-video/.

UITP. (2019). *Report—Mobility as a Service*. Retrieved from https://www.uitp.org/sites/default/files/cck-focus-papers-files/Report_MaaS_final.pdf.

Zipper, D. (2019, April). Cities can see where you're taking that scooter. *Slate*. Retrieved from https://slate.com/business/2019/04/scooter-data-cities-mds-uber-lyft-los-angeles.html.

Chapter 4
Policies to Decarbonize the Transport Sector

Abstract This chapter presents a range of policies that support an effective decarbonization of the transport sector, considering both the measures that favor a reduction and/or a modal shift of transport demand, and those that promote the spread of cleaner vehicles. The chapter also provide examples of best practices, i.e., cases in which local, national, or international authorities have attempted and been able to successfully implement these measures. Following analysis of individual policies, the chapter will look to take a more holistic approach to policy implementation. In this context, it will illustrate the governance levels that involve each singular policy, the taxation issue, potentially emerging risks, and the distributional effects deriving from the implementation of these policies.

4.1 Introduction

This chapter discusses the different policies that may be adopted to foster effective decarbonization of the transport sector. The policies are divided into two categories depending on whether they involve passenger or freight transport. Within these categories, each contains two subcategories based on whether the policies discussed:

- favor a reduction and/or a modal shift of transport demand; or,
- promote the spread of cleaner vehicles.

The chapter discusses policies by following a common pattern. First, it explains what the policy consists of and what it aims to achieve. Secondly, it enumerates the measures which characterize the policy. Thirdly, it provides examples of best practices, i.e., cases in which local, national, or international authorities have attempted and been able to successfully implement these measures.

Following analysis of individual policies, the chapter will look to take a more holistic approach to policy implementation. In this context, it will illustrate the governance levels that involve each singular policy, the taxation issue, potentially emerging risks, and the distributional effects deriving from the implementation of these policies.

Horizontal analysis shows clearly that in developing successful policies for the decarbonization of transport policymakers face a multitude of challenges. The first

© The Author(s) 2020
M. Noussan et al., *The Future of Transport Between Digitalization and Decarbonization*, SpringerBriefs in Energy,
https://doi.org/10.1007/978-3-030-37966-7_4

criterion for the assessment of a policy would naturally be its effectiveness in reducing carbon emission, in itself given the complexity of modern economic systems is not an easy task. Moreover, policymakers must consider the interaction of policies and their broader socioeconomic impacts. Considering effects on the price and efficiency of transport would be prime candidates here. More particularly, the regressive nature of many policies is an area that must be considered and addressed. Finally, many policies are predicated on the development of large-scale, sophisticated digital software and data processing. An investigation into possible side effects of digitalization is outside the scope of this chapter but their existence should be noted nonetheless.

4.2   Passenger Transport—Reduction and Modal Shift of Transport Demand

4.2.1   Promote the Increase in the Average Occupancy of Vehicles with Carpooling (Enabled by Digital Technologies)

Carpooling is a form of flexible transport whereby drivers of private cars offer their empty seats to other individual travellers that have similar itineraries and time schedules. There is little formal legislation controlling carpooling; however, it is common across the developed world that payments between passengers and drivers in a car pool should not result in a profit for the driver otherwise this could violate terms and conditions of their car insurance.

Carpooling schemes can be classified into two categories, which differ depending on travel's length: urban (or short-distance) carpooling and inter-city, inter-state (or long-distance) carpooling.

Starting from early 2000s, new companies launched long-distance carpooling services in Europe. The French BlaBlaCar is one of the most famous companies operating in this sector. This service has rapidly become popular and has been expanded in several countries. Long-distance carpooling is an almost entirely spontaneous phenomenon and therefore proper legislation to regulate it is missing. Similarly, there is no significant political initiative to encourage the spread of inter-city carpooling.

As regards short travels, carpooling options encourage car owners to cut urban congestion by inviting people to share a vehicle. Typically, such schemes are led by workplaces and include information technologies such as online booking. Dedicated infrastructure and specific policies may promote the uptake of urban carpooling. The main measures generally include:

- establishing a lane for high-occupancy vehicles on an access road into city centers;
- establishing park and pool venues on main roads;
- integrating carpooling with public transport;
- offering a trip-matching service for carpooling;
- carrying out the necessary public relations work and raising awareness.

A carpooling scheme has been tested in Rome (CIVITAS Initiative, Roma) in the framework of the European CIVITAS projects. During peak hours, many of the cars circulating in the center of Rome have a single occupant travelling to work. Some companies had carpooling "crews," but they tended to be formed spontaneously rather than in an organised way. The idea behind this measure was therefore to develop a systematic approach in order to promote collective car use more effectively.

The Sustainable Mobility Department of the public transport operator (ATAC) installed a simulation model designed to manage carpooling services and integrated it with the mobility management system. The model uses origin and destination data and incorporates revealed preferences about trip characteristics and scheduling.

In the first phase of the project, ATAC identified the needs and priorities of companies and individuals as well as the areas to promote carpooling in with the help of computer-based tools. After that, it cooperated with citizens to develop first home-to-work plans to serve as a trial. Finally, ATAC designed and promoted several carpooling schemes across the city.

The three main schemes involved the municipality's offices, the hospital Policlinico Umberto I and the Ministry of Public Health. Almost 400 car pools were formed and more than 1000 citizens took part in this experiment. Furthermore, the projects included agreements with local garages to provide facilities at special rates.

Stuttgart also provides a successful experience of implementation of carpooling systems (CIVITAS Initiative, upgrading the carpooling system with an events feature). The Pendlernetz Stuttgart carpooling system offers all commuters a chance to find an appropriate car pool, enabling them to travel to work in a more relaxed and environmentally friendly way. It was created in order to improve the mobility of all citizens and visitors to the city and the Stuttgart region. Digital technologies also improved the system: communication on potential car pools via mobile phone text messages to users, geographically referenced route mapping and automatic transfer to the public transport information system contributed to providing an effective carpooling service.

The carpooling system was extended to local events like football matches and concerts. This involved integrating an events data pool into the Pendlernetz Stuttgart system. The new service was aimed in particular at visitors to the home games of the football club VfB Stuttgart. During such events the demand for travel to the same destination (e.g., the stadium) skyrockets. A carpool scheme, therefore, proves even more efficient in such situations.

Other interesting examples in USA relate to urban carpooling. An MIT study developed an algorithm finding that just 3000 four-passenger cars could serve 98% of taxi demand in New York City (NYC), with an average wait time of only 2.7 min. Compared with the 14,000 taxis currently operating in NYC, the cost and emissions savings from carpooling are self-evident. The research does however highlight the benefits and dependency of successful carpooling on sophisticated digital techniques.

Recently, in Abu Dhabi the Department of Transport has for the first time embraced car sharing. They have launched a new online system encouraging motorists to sign up and offer lifts to strangers. Individuals signing up can either look for rides or offer them themselves. Charging for these rides will be illegal outside of sharing the cost of fuel and other expenses. The opportunity for carpooling has now been incorporated into the official transport website alongside other modes of transport such as cars, ferries, and traditional public transport (Integrated Transport Center). Importantly, the scheme has been launched in combination with the introduction of new toll gates on routes into and out of Abu Dhabi Island. The combination of these two strategies is targeted at reducing the number of cars on the roads (The National, 2019).

Efficient carpooling relies on a well-functioning platform. If carpooling is considered as an attractive option by cities, they should first look to best-case examples, such as those from Rome and Stuttgart to benefit from existing experiences. Trials should then be carefully rolled out, with adaptions made as they are increased in size.

4.2.2   Promote a Shift from Private Cars to Public Transport and Clean Car Sharing (Enabled by Digital Technologies)

This section will first examine policies to bolster a modal shift toward public transport. With this respect, it will describe the experience of the city of Usti nad Labem. After that, the section will focus on the shift toward a more modern transport mode— car sharing—by looking at the development of this new service in both American and European cities.

With the aim of influencing the modal shift in favor of public transportation, it is necessary to advertise the benefits and the services of this transport mode. Public transport promotion aims at supporting its good image, attracting passengers, and strengthening its usage. An effective cooperation between the city and its public transport operator is a prerequisite for a successful promotion campaign.

The basic objectives of public transport (PT) promotion includes:

- Highlighting advantages of public transport compared to individual motor transport;
- Attracting new passengers while keeping the existing ones;
- Improving provision of information about transport services;
- Increasing awareness about public transport and ITS;
- Improving accessibility of services by providing targeted information to passengers; and
- Exploiting new technology to supply favorable services, such as SMS ticket, downloadable public transport maps or timetable.

The city of Usti nad Labem in Czech Republic implemented a public campaign promoting public transport in the city to raise awareness of passengers and build reputation of public transport (CIVITAS Initiative, Public Transport Promotion Campaign).

The authorities produced and distributed relevant information, promotional, and education materials. Further distribution of these materials was transmitted to the PT Company of Ústí nad Labem, which will continue in the initiated activities. By providing clear information, the city and the relative transportation operators should facilitate the accessibility to public transport services for existing or potential users.

The campaign consisted of several events organized to promote local public transport services to residents. During these events, people were able to compare PT services provided in the past and in the present. They participated in knowledge and effort competitions for prizes, discussions, and workshops about PT services. Information materials were distributed during these events and they are available at other public areas and through the PT Company of Usti nad Labem. Two PT vehicles were equipped with free internet connection for passengers and decorated by thematic pictures in order to attract attention and welcome customers.

Local media also played an important role in promoting the campaign activities. Particular attention of the promotion campaign was laid on education of children to utilise PT in the city.

As regards results, the usage of the city public transport in Ústí nad Labem is now relatively high compared to similar cities in the Czech Republic, although improvements of services are still needed.

Shared mobility emerged in major cities in the early 2000s with the arrival of car sharing services from providers such as Zipcar, whereby members could borrow cars on a short-term (hourly) basis. Many of these services are now accessed through smartphone apps, allowing users to locate and unlock vehicles, and making one-way journeys possible.

Several cities have already allowed and pushed private companies to deploy electric vehicles in their urban car sharing networks. Zipcar in Chicago has used electric vehicles in its fleet since 2012. Philadelphia and Portland also are working with Zipcar on their programs. Other examples include Houston's Fleet Share, San Diego's SmartCitySD with Car2go, and BMW's program in San Francisco (Lutsey, Searle, Chambliss, & Bandivadekar, 2015).

In Italy, many communities and regions participated in the founding of their regional car sharing providers. This was pushed forward by the national coordination point for the development of car sharing, the Iniziativa Car Sharing, and its support through the Italian Environment Ministry. In several cities, communities are directly involved in the regional car sharing organization. Political support can be seen in the preferential treatment given to car sharing vehicles in many Italian cities:

- They have unrestricted access to the low emission zones in city centers—established because of poor air quality levels. Regular car traffic may enter only within given time limitations.
- Car sharing vehicles may use reserved bus lanes, bringing them through the worst congestion areas of the city more easily.
- They can park free of charge in the "blue zones" of the city center. Examples of such political support through local transport policy are found in Turin, Venice,

Bologna, Rimini and Modena. In addition, the councils in many Italian cities use car sharing for their employees' business travel.

4.2.3 Promote a Shift from Private Cars to Rail in Long-Distance Travels (E.g., TEN-T)

In order to foster a modal shift from road to rail, customer and market oriented measures as well as enhanced rail systems are required.

Measures to increase the capacity of the rail system include[1]:

- Investing in high-speed rail.
- Improving timetable planning: "e.g. double track—bundling of trains with the same average speed in the timetable channels and daytime operation of faster freight trains (Islam, Ricci, & Nelldal, 2016)."
- Exploiting trains with higher capacity.
- Differentiating track access charges to avoid overloaded links.

The EU has elaborated the 4th Railway Package (European Commission, Mobility and Transport), a set of six legislative texts designed to complete the single market for rail services (Single European Railway Area). Its overarching goal is to revitalize the rail sector and make it more competitive vis-à-vis other modes of transport. This package comprises two "pillars": the market and the technical pillar.

The market pillar includes a set of rules that allow railway undertakings from one Member State to provide passenger services in any other EU Country. It also guarantees impartiality, prevents discrimination, and introduces the principle of mandatory tendering for public service contracts in rail. As a result of these measures, the passenger rail sector will become more competitive. Increased competitiveness will stimulate operators to be more responsive to customer needs, provide better quality services and improve their cost-effectiveness. At the same time, the principle of mandatory tendering for public service will lead to savings of public money.

The second pillar, the technical one, provides a legislative framework enabling a considerable cut of the operational costs and administrative burden. In so doing, it boosts the competitiveness of the railway sector. In particular, it will:

- "Save firms from having to file costly multiple applications in the case of operations beyond one single Member State. ERA will issue vehicle authorizations for placing on the market and safety certificates for railway undertakings, valid throughout the EU. So far, railway undertakings and manufacturers needed to be certified separately by each relevant national safety authority.
- Create a "One stop shop" which will act as a single entry point for all such applications, using easy, transparent and consistent procedures.

---

[1]For further details on this topic see Islam et al. (2016).

- Ensure that European Rail Traffic Management System (ERTMS) equipment is interoperable.
- Reduce the large number of remaining national rules, which create a risk of insufficient transparency and disguised discrimination of new operators (European Union Agency for Railway, The 4th Railway Package—What does it mean for me?)."

In China, the central government is aiming to integrate AI technologies and its national railway to monitor and maintain its operations. Due to its limited expertise and experience, China Railway cooperated with the technology company Tencent and the automotive group Geely to install in-cabin WiFi on its trains. Incorporating social capital is seen as a move to rapidly develop China Railway's services.

China has also invested enormous physical capital into the development of the world's largest high-speed rail service. Work began in August 2008 with a line built connecting Beijing and Tianjin in just 30 min. Since then, China has put into operation over 25,000 km of dedicated HSR lines. These projects have been overseen by the 2004 Medium- and Long-Term Railway Plan covering freight and passenger transportation till 2020. Strong commitments from top levels of governance have been key for the success of HSR. There have been clear positive network externalities as HSR has sprawled across Chinese landmass. For instances until 2012, the Zhengzhou-Xi'an line was an isolated service until 2012. Following its connection to the Beijing–Guangzhou HSR in 2013, passenger volume increased by 43% (World Bank Group, 2019a). Extensive planning in China has driven an increase in passengers travelling via rail.

A clear priority for neighboring countries or states with land borders should be greater coordination between train services and investment in order to better stimulate the development of more efficient and high-speed rail. In terms of high-speed rail, lessons can be learned from the relatively successful story in China so far. A key criterion has been the investment of large amounts of capital. Countries should look for effective strategies of sharing this investment burden with the private sector.

4.2.4 Develop Cross Border Corridors for Connected and Automated Mobility (CAM)

New digital technologies have been increasingly applied to the transport sector. Smarter and more connected vehicles are currently under development. These new generation vehicles and infrastructure perform more efficiently by collecting large amounts of data. In particular, they are able to interact with other smart systems and optimize their performance, thus reducing energy consumption and maintenance costs.

This development involves different modes of transport (IEA, 2017a). In road transport, digitalization enables coordinated fleet operations as well as other operations such as the possibility to check the status of battery charging for electric vehicle. In aviation, modern sensors applied on new commercial aircraft and big data analytic systems enhance route planning and help rationalize fuel consumption. Maritime transport also benefits from improved communication systems that ships use to exchange information with port authorities. By doing so, operators prevent congestion in ports.

Automated mobility has also made concrete steps forward. By replacing human control, automated cars can improve safety while optimizing speed and gasoline (or energy) consumption. Major progress is also expected in the passenger railway and aviation.

Given the huge amount of data that need to be processed, the development of connected and automated mobility will be favored and enabled by the deployment of the 5G technology. To this end, the EU is establishing a collaborative network of cross-border corridors between Member States.

As part of the European Commission's 5G Public Private Partnership, the EU supports three 5G cross-border corridor projects for large-scale testing of connected and automated mobility (European Commission, 5G Public Private Partnership, the Next Generation of Broadband Infrastructure). The three projects, launched in November 2018, trial 5G technology applied to CAM over more than one thousand kilometers of highways across four borders:

- 5G-CARMEN (5G Carmen): 600 km of roads across an important north–south corridor from Bologna to Munich via the Brenner Pass.
- 5GCROCO (5G Croco): over highways between Metz, Merzig and Luxembourg, crossing the borders of France, Germany, and Luxembourg.
- 5G-Mobix (5G PPP): along two cross-border corridors between Spain and Portugal, a short corridor between Greece and Turkey, and six national urban sites in Versailles (France), Berlin and Stuttgart (Germany), Eindhoven-Helmond (Netherlands) and Espoo (Finland).

China is seeing the growth of Intelligent and Connected Vehicles (ICVs) which are able to achieve the exchange and sharing of information while being aware of complex surroundings. Through intelligent decision-making such vehicles may carry out driving operations independently of human beings. The Chinese Government has taken concrete steps to foster the creation of an intelligent transport eco-system. In 2017, a National Innovation Platform for the Acceleration of ICV Development was established. This platform looks to address obstacles in the development of ICVs as well as ensure the effective implementation of national strategies (GIZ, 2018).

In Shenzhen, the municipal government is participating in a public–private partnership called the Greater Bay Area Intelligent Vehicle Eco-Partnership (GIVE). The objective is to accelerate the formation of this ICV ecosystem around Hong Kong, Macau, and Guangdong province. Over 100 members from business and academia have joined as of 2018 (Ibid.).

R&D efforts should be continued and encouraged in order to develop more efficient connected and automated solutions.

4.2.5 Congestion Charging Policies

A congestion charge is a charge for driving a vehicle in an urban area, often limited to working hours. Devised as a way to internalize external costs, this initiative would be beneficial for two reasons. First, the congestion charge pushes up the cost of using private cars, thereby prompting users to choose other options. Reducing traffic in usually congested urban areas makes districts more enjoyable and attractive for potential new pedestrians or cyclists. Secondly, pricing congested road or areas raises money to build new cycle paths or make streets more suitable for walking.

Congestion charging schemes can be classified into four types (CURACAO, What Are The Key Features And Examples Of Scheme Design):

- Point based charges (e.g., tolls to cross a bridge or to enter a section of motorway).
- Cordon based pricing: A charge is levied for crossing a cordon, and may vary with time of day, direction of travel, vehicle type, and location on the cordon. There could be a number of cordons with different prices.
- Area license based pricing: A charge is levied for driving within an area during a period of time. The price may vary with time and vehicle type.
- Distance or time based pricing: Price is based upon the distance or time a vehicle travels along a congested route or in a specified area, and may vary with time, vehicle type, and location.

Congestion charging initiatives have been implemented in major cities. In London, this has resulted in a significant decrease (-21%) of congestion in the interested areas. As a consequence, the city reallocated more road space to sustainable transport modes. Walking and cycling also become safer as congestion drops: With this respect, a survey shows a significant fall in bicycle accidents in London (Guardian, 2015). CO_2 emission reductions of 12% were recorded, while a 12% reduction was also recorded in local air pollution, specifically PM10 (Joint Research Center, 2019, p. 451).

Similarly, in Milan private vehicles entering the charging zone have decreased by 29% since the establishment of the Area C, a combination of a congestion charge with a low-emission zone that cannot be entered by the most polluting vehicles. This congestion reduction has led to a 35% reduction in CO_2 emissions (Ibid.). At the same time, this initiative promoted a shift toward environmental friendly transport modes as the number of bike users significantly grew in the next three years (Polinomia srl, 2016).

In 1998, Singapore transitioned from a manual, labor-intensive pricing system to the current system of Electronic Road Pricing, becoming the first city in the world to do so. The system uses entry points to charge drivers entering the city center. Rates of entry vary depending on the time of day and between points of entry. All resident cars use in-vehicle stored-value smart cards while visitors must rent or purchase an in-vehicle unit. The novelty of such an electronic system is that, unlike a simple toll system, it directly charges drivers for the congestion they are causing though a variable price mechanism that increases during times and in areas of heaviest road usage (Menon & Guttikunda, 2010).

Cities should adopt more stringent congestion charging policies. The benefits must be carefully explained and discussed with the general public to avoid backlashes. A focus on discussing the health benefits that will arise through the reduction of local air pollution should be a fundamental mechanism through which to generate widespread support for higher levels of congestion charging. Particularly, older and heavily emitting vehicles should be targeted as a priority. Automatic and Electronic Road Pricing should be the preferred design, with Singapore providing a case study on best implementation.

4.2.6 Parking Management Policy in City Centers

At city level, one opportunity to reduce congestion and promote a modal shift of transport demand is the parking management policy. That is, local authorities could consider introducing or raising parking fees in certain districts. Through these kinds of policies, the costs connected to the use of private vehicles would increase. Higher, even though indirect, costs would discourage citizens from driving private means of transportation. The reduction in the private car usage should entail a rise in the public transport use (if this is present and efficient) or cycling (if bicycle paths are in place).

In 2012, Nottingham adopted a parking management policy to tackle problems associated with traffic congestion. In particular, the city introduced a Workplace Parking Levy (WPL) (Nottingham City Council). This refers to a charge on employers who provide workplace parking. WLP has provided funds that the city has invested in new transport infrastructure (such as the extensions to the existing tram system) and the redevelopment of Nottingham Station.

In India, New Delhi is currently undergoing an extensive overhaul of their parking management policy. Measures are aimed at curtailing the use of private vehicles, giving more space to pedestrians, and encouraging alternative, greener modes of transport (The Hindu, 2019). Policies include on-street parking charged at twice the value of off-street parking as well as exponentially increasing parking fees with time spent parked. Moreover, within each parking facility civic agencies shall identify and provide areas for electric vehicle charging and battery swapping facilities.

Draft rules of the New Delhi Parking plan (dated 12/06/19) detail that on-street spaces need to be utilized for the convenience of primarily pedestrians/cyclists, then mass public transport followed by emergency vehicles (Government of National Capital Territory of Delhi Transport Department, 2019). Civic agencies shall endeavor to use additional parking revenue for local development works related to the safety of pedestrians and road safety. If successfully implemented, the new parking management policy will engender a significant shift in population behavior away from individual car driving in one of the world's most populous cities.

4.2.7   Promote Cycling and Walking Zones

In many urban areas around the world, people still opt for traditional private cars because of the absence of alternative options. In this context, promoting cycling and walking zones would boost the modal shift of transport and ease traffic. That is, municipalities should make their urban spaces safer and walking or cycling friendly. To this end, they should prioritize the development or expansion of relevant infrastructure like cycle tracks and sidewalks. But also lower car speed limits and introduce car-free days.

As regards the promotion of cycling zones, Copenhagen's experience is surely among the best practices. In 2011, it developed its bicycle strategy (City of Copenhagen, 2011). The measures of the strategy included:

- Dedicating more space to cyclists: widening some tracks and building alternative routes to move some of the bicycle traffic away from the congested routes;
- Developing campaigns focused on more considerate behavior in traffic;
- Improving travel times by bicycle, for instance by prioritizing ambitious short cuts like tunnels and bridges over water, railways and large roads;
- Strengthening partnerships with companies, shopping districts, public transport providers, and neighboring municipalities to promote the use of bikes;
- Combining bicycles with public transport. Integrating bike sharing system with buses, trains and metro;
- Inviting bids for new bike sharing systems;
- Investing in new bicycle parking.

The adoption of this strategy has resulted in significant improvements. A 2017 report enumerates major achievements: 41% of all trips to work and study to/from Copenhagen was by bike and 62% of Copenhageners chose to bike to work and study in Copenhagen (City of Copenhagen, 2017). In total, 1.4 million km was cycled in the city on an average weekday which is an increase of 22% since 2006. In the same period, cyclists' feeling of safety has increased by 43% while the relative risk of having a serious bicycle accident has been reduced by 23%.

A case study in New Zealand has highlighted the potential of cycling and walking interventions to reduce carbon emissions (Keall et al., 2018). In 2010, New Plymouth and Hastings were selected by the New Zealand government as walking and cycling Model Communities. Total investment into the projects was $13.1 million of which 85% was spent on infrastructural changes (i.e., improved walkways and cycle lanes) and 15% on information and education (e.g., campaigns to increase cycling and walking uptake).

Based upon these interventions, a quasi-experimental study revealed that the average vehicle kilometers driven was reduced by 1.6% as an average of the two cities relative to control cities, similar in all characteristics other than not having enjoyed the investment. An estimated 1% reduction in annual CO_2 emissions follows as a result of this reduction in kilometers driven.

4.2.8 Promote Multimodality

As of today, most transportation systems in the world favor the choice of uni-modal solutions by users. This is explained by the fact that the concept of multimodality (or intermodality) is new and not yet widespread. Multimodality refers to seamless trip-making across multiple transport modes (e.g., walking, cycling, bus, urban rail, and cars) (IEA, 2017a, p. 40).

Most urban areas still lack effective schemes to integrate long and short-distance transportation means. Similarly, cities do not rely on frameworks to connect private vehicles on the one side and public or shared mobility services on the other. As a result, end users find single transport mode journeys more practical than multimodal solutions.

Multimodality can provide a large variety of benefits. It may contribute to resolve problems related to congested roads and offer cleaner and safer mobility opportunities. In fact, multimodality exploits the strengths of different transport modes, such as speed, cost, reliability, and convenience. By taking advantage from the combination of several mobility options, it provides transport solutions which are better suited to the needs of the user.

Measures aiming to develop efficient multimodal platforms include:

- Setting targets: what area must be connected to other areas;
- Defining the spatial scale: regional, national, or local level;
- Identifying transportation means to be connected: trains, private cars, buses, bikes, etc.;
- Setting a suitable, legal framework to facilitate the introduction of digital systems for passenger ticketing, real-time information sharing and reservation;
- Defining an investment plan to smooth interchanges at bus and train stations or other terminals;
- Involving private shared mobility providers and stakeholders interested in developing an interface to provide access to mobility options via a single ticketing and payment channel;
- Managing and regulating the charges for the use of interchange terminals.

The EU has been covering these themes and supporting projects aiming to promote multimodality. In particular, the project CLOSER identified new transportation patterns based on the idea of multimodal mobility to connect long and short-distance networks (European Commission, Project CLOSER). More recently, the EU has co-financed the CIVITAS projects that inter alia focus on topics like park and ride, public transport and bicycles, and the improvement of interchange terminals to promote intermodal urban transport chains at city level. For example, the city of Turku in Finland is improving the interconnectivity of its public transport system, bike sharing services, and private vehicles through the development of a participation platform using ICT and social networks (mobile apps, social media) (CIVITAS Initiative, Turku).

In the USA, collaboration between Uber and Rapid transit agencies such as the Metropolitan Atlanta Rapid Transit Authority (MARTA) aims to address this problem (Metro-magazine, 2015). MARTA app users are linked directly to the Uber app while in transit in order to help them reach their final destination. In turn, Uber drivers have information about when the bus or train will arrive so that they are already waiting at the station when passengers arrive.

Multimodal forms of transport are essential to combating passenger problems of the "last mile." That is, public transport is often very useful in transporting passengers from one city center to another or from one pre-defined station within a city to another but passengers face the difficult task of how to commute the "last mile" to their personal destination. In such a circumstance, they face an incentive to simply drive the whole route.

4.2.9 Limit the Number of License Plates to Be Registered Each Month

In large and highly polluted urban areas, improving alternative transportation options has not been successful in promoting a modal shift. Neither have policies adopted to discourage, through fiscal or non-fiscal means, users from driving their cars. In such cases, local authorities have considered a strict limitation in the number of license plates to be registered each month.

One of the most extreme examples is Beijing where tackling problems related to air pollution and traffic congestion are priorities. To this end, since 2011, the municipal government has devised a lottery to restrict the number of cars registered each year (Bloomberg, 2019b). This policy is so strict that just one new plate for every 2000 applicants is awarded bimonthly. In fact, over the last years, the annual new vehicle quota has fallen to 100,000, from 240,000 in 2013.

Also, the local authority requires each licensed gasoline-fueled car to be idle one day a week (the day is determined by its license-plate number).

Following the implementation of these restrictions, car sharing and ride-hailing companies such as Didi Chuxing and Shouqi Limousine & Chauffeur acquired considerable share in the transportation market. The strong impact of these policies has even changed the consumer attitude from necessarily owning a car to booking a car on demand.

Placing an explicit limitation on car registrations is a reasonably extreme measure. However, it is likely to become more necessary in the future unless governments begin to quickly and effectively adopt many of the other policies detailed in this chapter.

4.3   Clean (Automated and Connected) Vehicles for Passengers

4.3.1   Emissions Standards

One of the most common policies designed to cut CO_2 emissions in the transport sector consists of setting emission standards. This kind of policy generally identifies a certain level of emissions (measured in CO_2 grams per kilometer) for newly registered vehicles. In the event in which these targets are not met the legislation usually stipulates financial penalties. For example, if the average CO_2 emissions of a certain manufacturer's fleet exceed the target defined by the authorities in a given year, the manufacturer has to pay an excess emissions premium for each car registered. Also, emission standards provisions are often accompanied by directives on fuel quality, which require a reduction of the greenhouse gas intensity of transport fuels.

In China, the corporate average fuel consumption limit for passenger cars is dropping from 8.2 l/100 km in 2010 to 5.0 l/100 km in 2020, and then 4.0 l/100 km in 2025 (Lubrizol, 2019). According to estimations (Ibid.), in 2020, there will be a saving of around 160 billion liters of fuel (over a trillion RMB) and 370 million tons of CO_2 every year.

In 2017, the EU average emissions level of the new cars registered was 118.5 grams of CO_2 per kilometer (g/km) (European Commission, reducing CO_2 emissions from passenger cars). From 2021, phased in from 2020, the EU fleet-wide average emission target for new cars will be 95 g CO_2/km.

Within its Mid-Century Strategy for Deep Decarbonization, the USA also established fuel economy and GHG emissions standards for both light- and heavy-duty vehicles, which have reduced transportation GHG emissions significantly in recent years (US Mid-Century Strategy for DEEP DECARBONISATION). However, on April 2, 2018 the Trump Administration announced its intent to revise through rulemaking the federal standards that regulate fuel economy and greenhouse gas emissions from new passenger cars and light trucks.

Emissions standards are, by definition, an effective tool for reducing emissions associated with road transport. Governments should communicate clear plans for progressively increasing the stringency of emissions standards over time.

4.3.2   Country-Level Bans on Commercialization of Petrol/Diesel Cars

In order to meet emission reduction targets, governments have increasingly considered policies to promote the use of zero emission vehicles. Aiming to boost the commercialization of cars and vans powered by clean energy, some countries have committed themselves to end the sale of new conventional petrol and diesel vehicles in the future. In so doing, they expect almost every car and van in circulation 10–15 years after the introduction of the ban to be zero emission.

With respect to this policy, European countries seem to be at the forefront. Denmark has proposed a ban on the sale of new cars with internal combustion engines from 2030 and hybrid from 2035 (Reuters, 2018). Prior to Denmark, France and the UK had pledged to stop supplying fossil fuel-powered vehicles by 2040. The UK government has subsequently proposed to bring this target forward to 2035 (BBC, 2020). The Dutch Government has proposed a plan to ban all petrol and diesel vehicles by 2030, so that all new cars in the country must be emission-free (Wiredbugs).

India has recently announced that its commercialization of petrol and diesel cars will cease in 2030. China has also joined the initiative and announced that its ban will cover the commercialization of internal combustion engine cars and commence in 2040.

Country-level bans can be viewed as an effective way to shift the incentives for car manufacturers toward producing non-petrol or diesel vehicles. Bans are in that sense very attractive but must be accompanied by more intermediary measures to help guide developments toward a longer-term target. Bans will only serve to penalize the population if there are not already feasible transportation alternatives in place.

4.3.3 Public Investments in Clean Vehicles R&D

Driven by climate change and air pollution concerns, research and development of electric and other clean vehicles have become a crucial tendency in the transportation sector.

China is the country which has allocated most financial resources to clean vehicles R&D.

China began R&D for electric vehicle under the National High-Tech Development Program "863 Program" (Report on National R&D Programmes on the Fully Electric Vehicle). The national 863 Program was introduced during the 10th Five-Year Plan (2001–2005), where also the goal to commercialize and industrialize electric vehicles was introduced. This goal was refined in the 11th Five-Year Plan (2006–2010).

The main objectives of the "863 Program" are to fund technological research and innovation in areas of strategic importance to the nation's economic and social development. In particular, it is engaged in the funding of electric vehicle related activities. To this end, in 2001, the "863 ElectricDrive Fuel Cell Vehicle Project" received an initial investment of RMB 800 million (approximately equivalent to 103 million Euros).

This strong public engagement has resulted in significant development of EV technology, which, in turn, has reduced EV production costs, in particular by reducing battery costs, further nurturing the development of the EV market.

The EU too has made concrete steps by adopting the strategic action plan for batteries. In particular, the European Commission:

- In collaboration with Member States makes available, research and innovation funds (H202018) for battery-related innovation projects, according to pre-identified short- and longer-term research priorities across the batteries value chain.
- Launches calls for proposals for an additional total amount of EUR 110 million for battery-related research and innovation projects (in addition to EUR 250 million already allocated to batteries under Horizon 2020; and EUR 270 million to be allocated in support of smart grids and energy storage projects as announced in the Clean Energy for all European package).
- Supports the creation of a new European Technology and Innovation Platform to advance on battery research priorities, define long-term visions, and elaborate a strategic research agenda and road-maps.
- Prepares the launch of a large-scale Future Emerging Technologies Flagship research initiative, which could support long-term research in advanced battery technologies for the 2025+ timeframe.
- Optimises solutions for integration of stationary storage and electric vehicles in the grid within Horizon 2020 smart grid and storage projects23 as well as Smart Cities and Communities' projects.

Alongside penalties and bans for the use of polluting vehicles, it is very important for governments to play a role in driving forward research into the cleaner technologies that will one day offer a replacement. The benefits from R&D and innovation in clean vehicles are likely to be far-ranging. Government schemes should look to effectively unlock private investment at the same time as providing some public funds.

4.3.4 Public Investments in Clean Vehicles Infrastructure (E.g., EVs Charging Network)

A successful strategy aiming to expand the electric vehicle ownership requires considerable investment in relevant infrastructure. Drivers must have confidence in the performance and connectedness of charging and fueling networks before they will purchase or use a zero-emissions vehicle (ZEV). With this respect, local, state, and utility stakeholders need to cooperate to stimulate the accessibility and ease of use of all types of charging infrastructure, both public and private ones.

The State of California has invested much effort on the development of an effective charging infrastructure. In January 2018, the Californian government signed an Executive Order, setting ambitious targets of 200 hydrogen fueling stations and 250,000 electric vehicle chargers to support 1.5 million ZEVs on California roads by 2025, on the path to 5 million ZEVs by 2030 (Governor Edmund G. Brown Jr, 2018).

The initiative is designed to focus multistakeholder efforts on deploying charging and fueling infrastructure as well as making ZEVs increasingly affordable to own and operate.

In particular, the California Public Utilities Commission has approved roughly $1 billion in programs to incentivize residential, workplace, and public charging from Pacific Gas & Electric, Southern California Edison and San Diego Gas & Electric (Greentech Media, 2018).

The intent of the government is also to help expand private investment in zero-emission vehicle infrastructure, particularly in low income and disadvantaged communities. With this in mind, the 2018 Executive Order serves inter alia the purpose to develop a concise document mapping the relationships between existing and planned ZEV infrastructure investments. Moreover, it identifies the need to create a platform for stakeholder engagement, feedback, and information sharing.

The Chinese central government is also aggressively pursuing the development of EV charging networks. As of January 2019, there were roughly 2.6 million EVs on the road in China. There were 808,000 eV chargers, of these roughly 330,000 were public and 480,000 were home chargers. Charging infrastructure is growing rapidly throughout China, with major cities (Beijing, Shanghai, Guangdong Province) having taken the lead on this front. The majority of provinces in China added over 1000 new charging posts in 2018 alone. Highway corridors have also been created for EV charging between Beijing and Shanghai as well as other major cities (Center on Global Energy Policy, 2019).

Central government policies are clear on the drive for larger infrastructure. In September 2015, the State Council issued the "Guidance on Accelerating the Construction of Electric Vehicle Charging Infrastructure." This guidance called for a charging infrastructure sufficient for 5 million EVs to be developed by 2020. Furthermore, 10% of parking spaces in large public buildings must be available for EV charging (Ibid.).

There is a pivotal role for governments to play in providing the public good of charging networks. Incentives are reduced for both manufacturers to design, and consumers to purchase, clean vehicles without clear plans for an effective network in place. A structured plan would provide the necessary certainty to companies that investing in ZEVs will be worthwhile.

4.3.5 Clean Vehicles Production Quotas for Carmakers

Quotas usually refer to restrictive practices. For example, they may be used to limit imports thereby benefiting local producers. However, quotas may also serve the purpose to stimulate production instead of restricting it. More precisely, they may be devised to encourage the production of certain goods and curb output of others. In the context of the transportation sector, the goal of this strategy is to promote a shift. In fact, authorities may apply quotas on EVs production or import to car manufacturers. In other words, car industry corporations are obliged to reach a certain level of

EVs as a percentage of their total vehicle production or import. By encouraging the commercialization of electric cars, this kind of quota clearly boosts a shift toward cleaner vehicles.

The State of California adopted a quota policy to meet its health based air quality standards and greenhouse gas emission reduction goals. Some years later, China introduced a slightly modified version of California's program (International Council on Clean Transportation, 2018). The Chinese regulation was formally introduced in 2018 and applies only to passenger cars. It establishes New Energy Vehicle (NEV) credits, which need to be achieved by producing or importing enough new energy passenger cars. Higher performance vehicles get more credits. At the same time, the rule allows manufacturers to use surplus NEV credits to offset corporate average fuel consumption credit deficits.

The Chinese government specified EV targets for car manufacturers: 10% of the conventional passenger vehicle market in 2019 and 12% in 2020. Failure to meet these credit targets after adopting all possible compliance pathways will lead to Ministry's denial of type approval for new models that do not meet their specific fuel consumption standards until those deficits are fully offset.

Clean vehicle quotas can be an effective tool for encouraging production; however, governments should be careful not to create perverse incentives toward particular models and technologies. Any quota should remain as technologically neutral as is possible.

4.3.6 Public Procurement for Clean Vehicles

The market for clean vehicles may also be stimulated by direct purchase from the public administration. Although the means of transport used by the public services account for just a small share of the total vehicles in circulation, this measure may still impact on the deployment of low or zero emission transportation.

Authorities at regional or state level may commit themselves to purchase environmental friendly automobiles to raise the demand for this kind of vehicles. Also, and more importantly, they can promote clean mobility solutions in public procurement tenders by defining targets that cities and other authorities at local level have to meet.

The European Commission proposed a directive which sets minimum procurement targets for each category of vehicle and for each Member State (European Parliament, 2019a). The scope of the directive is broadened to include forms of procurement other than purchase, namely vehicle lease, rent or hire-purchase, and to public service contracts for passenger transport by road and rail, special-purpose road transport passenger services, non-scheduled passenger transport and hire of buses and coaches with driver.

The proposal sets minimum procurement targets for each category of vehicle and for each Member State. For light-duty vehicles, Member States must reach a share ranging from 16 to 35%, which is the same for the 2025 and for the 2030 deadlines.

For buses, individual Member State targets range from 29 to 50% (2025) and from 43 to 75% (2030), and for trucks from 6 to 10% (2025) and from 7 to 15% (2030).

Furthermore, the proposal introduces reporting and monitoring obligations for the Member States and aligns the Commission's and Member States' reporting obligations. It provides for intermediate reporting in 2023 and full reporting in 2026 on the implementation of the targets for 2025, and further reporting every three years thereafter.

In China, individual provinces have embarked upon their own public procurement of electric vehicles, backed by the central government with an ambition for China to establish a position as a world leader in battery technology. In 2017, Shenzhen became the first city in the world to realize the full electrification of its public bus fleet with a total of 16,359 electric buses. This was thanks to a combination of subsidies from central and local government. As of January 2019, 99% of taxis are also electric (The Guardian, 2018; Institute for Transportation and Development Policy).

The lessons learned from China and in particular Shenzhen are that public procurement can play a very effective role in stimulating the clean vehicle industry.

4.3.7 Subsidies and Other Special Provisions (E.g., Grants, Tax Credits, Tax Exemptions)

Clean vehicle commercialization can be boosted by fiscal incentives. Consumer purchasing incentives for EVs, just as subsidies on other goods, promote EV sale by reducing the final cost for consumers. Several states across the USA cover part of the EV costs. Equally common are subsidies for home chargers and public chargers in the form of tax credits, rebates, and grants.

Some states offer the same subsidies to all types of electric vehicles, some provide a different amount to plug-in hybrid EVs (PHEVs) and battery EVs (BEVs), and others offer the benefit only to BEVs (Lutsey et al., 2015). "Examples are California's Clean Vehicle Rebate Project, which offered $2500 for BEV and $1500 for PHEV purchases in 2013; Colorado's motor vehicle credit that offers up to $6000 based on battery capacity and purchase year; Massachusetts' MOR-EV rebates of up to $2500 for plug-in vehicle purchases; and Georgia's income tax credit for ZEV purchases and leases of 20% of the vehicle cost, up to $5000. Georgia's program is set to expire in mid-2015" (Lutsey et al., 2015, p. 15).

In some cases, income tax credits are accompanied by state sales tax exemptions for electric vehicle purchases and related services. Examples include Washington's retail sales tax exemption for alternative fuel vehicles; Maryland's excise tax credit of up to $3000 based on battery capacity for purchase or lease of a plug-in vehicle, and the district of Columbia's excise tax exemption for high fuel economy vehicles.

Aside from federal and state vehicle purchasing support, several cities have contributed to promoting electric vehicle purchases by offering additional financial support. Furthermore, utilities may offer a multitude of incentive programs for customers who own electric vehicles. Some utilities offer discounted rates for electric vehicle charging, and many offer time-of-use rates, which allow charging at much lower cost during off-peak hours.

It is important, in areas such as subsidies, that governments follow a clear and coherent plan. Stop-and-start subsidy programs, such as those typically experienced for household renewable energy subsidies, will not be effective in facilitating the development of a strong zero-emission vehicle industry that can survive when subsidies are eventually removed.

4.3.8 Non-fiscal Incentives (E.g., Parking Benefits)

Beside subsidies, non-fiscal incentives have played a significant role in promoting low emissions vehicles over the last years. The most common measures concern parking benefits and unrestricted access to high-occupancy vehicle (HOV) or carpool lanes for electric vehicle drivers.

Two US states provide free parking for electric vehicles. Hawaii offers free parking for electric vehicles at eligible parking locations that are metered (Slowik & Lutsey, 2017, p. 5). In Nevada, local authorities with public metered parking areas are required to launch programs for alternative fuel vehicles to park in these areas without paying a fee.

Cities also commit to provide parking benefits, New York City (NYC) and Denver being two examples. NYC demands that 25% of new parking be electric vehicle ready. Denver set a rule that calls for new lots with at least 100 spaces to have at least one designated for electric vehicles (Lutsey et al., 2015, p. 11).

As for the other measure, ten states offer unrestricted access to HOV or carpool lanes for electric vehicle drivers. California and Florida also exempt electric vehicles from toll charges on high occupancy toll (HOT) lanes, sometimes called "express lanes" but essentially HOV lanes that single occupancy vehicle drivers can access by paying a toll. Access to HOV and HOT lanes reduces the time that electric vehicle drivers spend on the road during peak traffic hours.

In China, the registration lottery for new cars is significantly more lenient and accessible for those who purchase an electric car compared with those who purchase a gasoline or diesel powered vehicle. Stringent restrictions and years long waiting lists for registrations of gasoline/diesel vehicles are targeted at encouraging reduced demand for driving alongside shifting some of this demand toward electric vehicles. Registered cars in Beijing must also lie idle for one day of the week determined by license plate number.

Governments therefore have many options, both fiscal and non-fiscal, for promoting the spread of low emissions vehicles. Most effective strategies should involve a combination of both.

4.3.9 City-Level Bans on Circulation of Petrol/Diesel Cars

As a part of a drive to clean up air pollution, some local authorities have considered significantly reducing CO_2 emissions by banning petrol or diesel cars from driving in the city.

To this end, the city of Amsterdam has drafted a strategy divided into three steps (Guardian, 2019). From 2020, diesel cars that are 15 years or older will be banned from going within the A10 ring road around the Dutch capital. From 2022, public buses and coaches that emit exhaust fumes will no longer enter the city center. The city council plans to ban cars and motorbikes running on petrol or diesel from driving in Amsterdam from 2030 completely. At the same time, the city is investing in charging stations and new infrastructure to promote the uptake of zero-emission vehicles and encourage citizens to switch their transport mode.

Meanwhile, Madrid has also begun to restrict car access to the city center. By 2020, all older diesel and gas-powered cars will not be allowed to enter the center at all. Hybrid vehicles with an "eco-label" will be permitted to enter.

Since January 1, 2018, Brussels has been a low-emission zone. Older diesel vehicles are now banned from the city center. The ban will be gradually extended to other diesel vehicles by 2025. The same applies to petrol vehicles: some grades are already banned while petrol vehicles up to the EURO 2 standard will be banned from 2025 (Brussels Capital Region).

In December 2016, mayors from Paris, Madrid, Athens, and Mexico City simultaneously announced plans to ban diesel cars and roads from their roads by 2025. It is not yet clear whether diesel vehicles in the four cities will be subject to a total ban, or which areas of the cities will be covered (The Guardian, 2016).

4.4 Freight Transport—Switch from Road to Rail

Freight transport represents almost one third of total transport emissions. This major impact is due to the large use of road transport which is responsible for the greatest part of the entire sector emissions. In fact, not only goods shipment by road accounts for the largest part of total freight sector, but it causes considerably more pollution than other transportation systems, such as railway. Therefore a switch from road to rail would have a significant, positive impact on the environment. Policies to support such a switch include the measures described below.

4.4.1 Subsidies

An increase of the modal share of rail freight can largely absorb the expected growth in freight transport and minimize its climate impact on the environment. A shift toward cleaner shipment schemes may be conducted by modernizing the rail freight industry, enhancing efficiency and standardization, improving cost-effectiveness and accelerating technological innovation. These aims may, in turn, be achieved through government financial support to rail freight operators. Benefiting from subsidy policies, operators will be expected to pass on their savings to their freight shipping customers via lowered rates.

In 2018, the German government announced that it had adopted an aid scheme to support a shift from road freight transport to rail transport (European Commission, State Aid). The scheme has a yearly budget of €350 million and will run until 2023.

In particular, the newly introduced policy provides that rail freight operators will be compensated for up to 45% of their track access charges, i.e., the charges that railway undertakings have to pay for the use of the rail network. By lowering operational costs faced by rail freight operators, this aid scheme makes rail transport more competitive in the shipment sector. Customers have, thus, incentive to shift their preferences from trucks toward a cleaner transportation system.

This kind of policy not only will benefit environment—as carbon emissions will be significantly reduced: it will also positively impact on congestion as medium and heavy freight vehicles' presence on the roads would decline.

In South Africa, the Department for Environmental Affairs in 2014 commissioned a report into the possibilities of a freight shift from road to rail. Since deregulation of the railways in the 1980s, the freight modal share by rail has been decreasing. The conclusion of the report was that such a shift is highlighted as one of the most favorable measures in reducing greenhouse gas emissions. One concern of the report was that rail freight is less labor intensive than road freight and so the transition could be associated with job losses (Department of Environmental Affairs RSA, 2014).

The Department of Transport has since released a Draft White Paper in which it details plans for rail to serve as the national land transport backbone by 2050 (Department of Transport RSA, 2017). Interventions will be two-pronged: infrastructure investment interventions to enhance rail's competitiveness, and enabling interventions to adjust institutional arrangements to ensure that rail functions effectively in delivering its share of national transport. It is recognized that funding of freight rail is inadequate and the government should ensure that additional sources are tapped. The government will limit its funding contribution to rail infrastructure only. This will consist of capital grants from national government as well as capital grants or long term investment instruments from provinces.

4.4.2   Cross-Border International Railway Connections

As of today, international goods shipment is mainly carried out by road. One reason explaining the lower use of rail is the lack of developed international infrastructure connections and service provisions. In order to promote a shift toward more environmentally friendly freight transport schemes, national authorities should coordinate themselves or with supranational institutions to improve cross-border international railway connections.

With this aim in mind, policies should consider the following measures:

- Implementing a stable long-term planning and financing framework for cross-border railway projects;
- Fostering coordination of national or supranational statistical offices to collect data and produce reports regarding the development of cross-border missing links;
- Funding cross-border infrastructure projects;
- Creating a platform to coordinate cross-border projects;
- Involving people at local level. Special attention must be given to resident's concerns about environmental impact and citizen's questions about the use of public money;
- Making information on cross-border rail connections available to potential customers operating in the freight segment;
- Providing a common, transparent and solid regulatory framework for infrastructure managing;
- Equalizing diverging, national taxes, administrative costs, and infrastructure access charges.

In order to develop a Europe-wide infrastructure network, the EU has established the Trans-European Transport Network (TEN-T), a policy which includes the implementation of cross-border railway lines.

The ultimate objective of TEN-T (European Commission, TEN-T) is to close gaps, remove bottlenecks and eliminate technical barriers that exist between the transport networks of EU Member States, strengthening the social, economic and territorial cohesion of the Union and contributing to the creation of a single European transport area (European Commission, Mobility and Transport).

The implementation of this policy is crucial to promote an effective modal shift in freight transport. This modal shift would result in an improvement of air quality. Preliminary estimates show a potential, significant impact on air pollution: the development of cross-border railway infrastructure, together with other measures carried out in the TEN-T framework, will boost the reduction in GHG emissions by about 7 million tons between 2015 and 2030 (European Commission—DG for Mobility and Transport, 2017).

In the USA in 1980, rail freight transportation accounted for just 20% of intercity freight miles down from 70% in the 1930s. The 1980 Staggers Rail Act deregulated many aspects of rail infrastructure development and service provision, allowing much greater flexibility in price setting and service levels. A stable environment encouraged investment, and between 1982 and 2009 the rail industry invested $510 billion in capital improvements. Technological standardization across the North American rail network played a crucial role in enabling long-haul freight transportation to be economical by rail. Since 1980, both the volume and productivity of US freight rail transportation have been steadily increasing (World Bank Group, 2019b).

While maintaining focus on improving national rail policy, governments must also recognize the importance of developing cross-border solutions if they are to develop an environment in which more significant volumes of freight transportation can be carried out via rail. Key policy considerations should revolve around stimulating coordination, standardization, and transparency for investors.

4.4.3   High-Speed Train

An effective policy designed to promote a modal shift toward greener transport modes may not only rely on the widening of the existing infrastructure. In fact, the enhancement of the service quality must accompany projects concerning the rail network extension. With this respect, investments on high-speed rail play a major role.

As of today, high-speed lines are generally utilized for passenger trains. This is due to the fact that the related infrastructure mostly reaches city centers which are generally destinations for passenger but not freight transport. Nevertheless, due to the current e-commerce boom, the demand for express delivery of goods, and thus high-speed freight transport, is fast increasing.

As a consequence of these new trends, there has been a growing interest from governments to improve freight rail which resulted in investments in high-speed infrastructure. In this context, Italy has been acting as a frontrunner. In 2018, the first national high-speed railway service became operational (RailFreight.com, 2018a). Operated by Mercitalia—the freight arm of Italian State Railway—this line will connect the terminal of Maddaloni-Marcianise in Caserta, the natural logistic gateway to Southern Italy, with the Bologna Interport, one of the most important logistics hubs in Northern Italy in three hours and thirty minutes. The goods will travel on board a high-speed train at an average speed of 180 kmph. The ETR 500 train has a load capacity equivalent to 18 tractor-trailers. According to the Italian service operator, "Mercitalia fast" is the world's first high-speed rail transport service for goods.

Although Mercitalia offers the first high-speed freight service in Italy, it is followed by another service planned to be launched in the same country (RailFreight.com, 2018b). In 2019, the Italian company Interporto Servizi Cargo (ISC) will offer a high-speed rail freight service between Florence and Bologna. Mercitalia and ISC will use the same high-speed railway line, but at different time slots.

Japan built its first high-speed rail link in 1964 connecting Tokyo to Osaka. Since then, Japan's regional structure with large metropolitan centers located a few hundred miles apart and a high demand for travel has favored the development of high-speed rail. The rail network has been significantly expanded geographically with current top speeds reaching 188 mph (302 km/h). The network has evolved to serve both freight and passengers; however, the HSR service is dominated by passenger services (Abalate & Bel, 2012).

Meanwhile, China currently owns the world's largest HSR system with more than 20,000 km of track in service. This network is also largely passenger dominated with HSR freight not as common. Surging demand for express delivery is slowly driving a shift. In 2013, China Railway Corporation implemented the first freight reform since its establishment by adding HSR freight. In 2014, China Railway Express began to provide HSR freight service in more than 100 cities. In October 2016, China Railway High-speed express service was launched offering customers door-to-door small parcel express services. The main good delivered were parcels, medicines, and emergency goods. This growth in express delivery may drive future development in larger-scale freight transport (Gao, Jiang, & Larson, 2017).

HSR travel in China is however still primarily designed for passenger traffic. HSR freight service mainly uses the vacant space in high-speed passenger trains. The operation time, section, and schedule are constrained by passenger traffic. Moreover, HSR passenger stations do not have sufficient freight operation conditions such as warehouses and stacking areas. The conditions in China are thus ripe for a burgeoning future of HSR freight transport but this has not yet been fully realized (Ibid.).

An obvious lesson is that any developments in HSR have so far been aimed at passengers while freight has been largely neglected. Under certain circumstances, there may be significant gains to be realized from stimulating a shift into HSR freight transportation. In countries that have already invested heavily into passenger HSR, there are likely to be substantial cross-benefits from a strategy that focuses upon utilizing HSR for both passengers and freight. Countries who have not yet significantly invested should consider optimal policies for combining passenger and freight HSR.

4.5 Cleaner and More Efficient Freight Transport

4.5.1 IMO Regulations to Reduce Sulfur Oxides Emissions from Ships

Beside policies aiming to promote a modal shift in the freight transport, significant efforts must also be devoted to making goods shipment systems more sustainable.

In the framework of the International Maritime Organization (IMO), governments have agreed on a regulation to limit air pollutants emissions from ships significantly. Ships are responsible for Sulphur oxides (SO_x) emissions, which are produced when crude oil is combusted in their engine. SO_x is known to be harmful to human health, causing respiratory issues and lung diseases.

The rationale behind the recently adopted regulation consists in a ceiling imposed to the sulphur content in fuel oil. Currently, for ships operating outside designated emission control areas, the limit for sulphur content of ships' fuel oil is 3.50% m/m (mass by mass). By contrast, under the new IMO's regulation, we are going to witness a substantial cut: from 3.50 to 0.50% m/m. The new limit is going to apply from January 1, 2020 (International Maritime Organization). Moreover a stricter, already in force ceiling of 0.10% m/m applies to four emission control areas: the Baltic Sea area; the North Sea area; the North American area; and the US Caribbean Sea area.

4.5.2 Encouraging the Use of LNG as a Marine Fuel (E.g., Rotterdam)

Research shows that the use of LNG as a marine fuel to replace current oil-based fuels has a significant, positive impact in terms of reduction of greenhouse gas emissions (Port of Rotterdam). Supporting the employment of LNG would thus contribute to making maritime transport more sustainable.

To this end, port authorities could commit themselves to the following initiatives:

- Providing land for LNG bunkering terminals, the construction of quay walls, jetties, or other possible basic infrastructure for maritime access;
- Playing a proactive coordinating role in conducting feasibility studies on LNG bunkering in cooperation with various stakeholders (i.e., local government, competent authorities, private actors, etc.);
- Adopting incentive policies to attract investments;
- Setting a differential port tariff on ships fueled by LNG or other clean fuels.

The Port of Rotterdam is at the forefront with respect to these practices. In particular, it has established a strategic partnership both with private stakeholders and another port in the region (Gothenburg) to develop LNG infrastructure and the associated safety and technical standards. In December 2018, Cees Boon, Senior Policy Advisor at the Harbormaster Policy Department of the Port of Rotterdam declared that by 2020, nine licenses will have been granted to LNG bunker providers to operate at the port (International Association of Ports and Harbors). Moreover, the Dutch authorities have considered a ban on dirty shipping fuel: the upcoming regulations (Bloomberg, 2019a) will bar most ships from using high-sulfur fuel oil which is essentially a by-product for refineries, thereby boosting the demand for LNG.

Policies carried out by the Dutch Port over the last years have proved successful. In 2018, the sale of bunker oil in the Rotterdam bunker port fell from 9.9 to 9.5 million m^3 (Safety4Sea, 2019). On the other hand, the throughput of LNG as bunker fuel increased considerably from 1500 to 9500 tons.

The government of South Korea has announced plans worth 2.48 billion USD to develop LNG bunkering facilities in the country (Maritime Executive, 2019). Singapore, as the world's largest bunkering hub, is also committed to the development of LNG as an alternate marine fuel. The Maritime and Port Authority (MPA) of Singapore is undertaking a variety of measures: co-funding of LNG vessels, waiving of craft dues of LNG-fueled craft, 10% port dues concessions for qualifying LNG vessels, and a pilot program testing the operational protocols of Singapore's LNG capabilities (Maritime and Port Authority of Singapore).

More particularly, the Singapore MPA will waive five years of craft dues for new LNG-fueled harbour crafts registering between Oct 1, 2017 and 31 Dec 31, 2019. An additional 10% of concessions in port dues will be granted to qualifying vessels that engage LNG-fueled harbor craft for their port operations. As of January 1, 2020, the sulfur content used aboard any ship docking in Singapore shall not exceed

0.50%m/m therefore indirectly increasing the demand for LNG fuel (Maritime and Port Authority of Singapore, 2019).

In October 2016, a Memorandum of Understanding was signed between Antwerp, Japan, Norway, Rotterdam, Singapore, and a range of other LNG-supporting port areas with the goal of deepening LNG bunkering cooperation and information sharing and promoting the adoption of LNG as a marine fuel (Maritime and Port Authority of Singapore).

Regulations on sulfur emissions are therefore an attractive strategy to directly reduce the release of harmful pollutants and indirectly stimulate the use of cleaner alternatives. Such regulation will be most effective when countries work together under international organizations, such as the IMO.

4.5.3 Supporting Truck Automation

Like passenger transport, the road freight sector is undergoing a process of automation enabled by new digital technologies. Digitalization has already paved the way for systemic improvements in the freight transport, the introduction of Global Positioning System (GPS) being a glaring example. Yet, new generation technologies have the potential to transform shipment activities fully, leading to the deployment of completely digitalized "driverless" trucks, automatized, and operated remotely.

Although vehicle automation has predominantly been associated to the passenger sector, there are good chances for this technology to penetrate the freight segment too for two reasons. First, automatized trucks would entail a considerable, operational cost reduction, since driver-based expenses would fall. Second, implementing this new technology in trucks would be easier as trucks mostly travel on highways. In fact, highways constitute a relatively predictable and stable driving environment compared with urban areas.

A successful process of gradual automation of road transport requires actions to be implemented both at local and international level. Local authorities are responsible for the so-called systemic improvements which include inter alia effective road maintenance. Beside the systemic improvements, authorities should set up an efficient environment in which a huge amount of digital data and information can be exchanged. With this in mind, we refer to the Sect. 1.4 "Develop cross border corridors for connected and automated mobility," which examines the EU's effort to establish a collaborative network of cross-border corridors between its Member States. This effort aims at developing the 5G technology to ensure an adapt environment for data processing.

In the USA, truck platooning refers to the concept of automated trucks driving sequentially and due to information sharing able to drive closer to each other than would be possible with human drivers. Such truck platooning has the potential to reduce fuel consumption by up to 10% while improving safety. Truck platooning has been successfully demonstrated on test vehicles by several developers in the USA. Regulatory proceedings are seen as the most prohibitive factor to more widespread technological roll-out. Some federal oversight is provided by organizations such as the National Highway Traffic Safety Administration (NHTSA) but a large proportion of relevant legislation is decided at the state level (New York Department of Transportation, 2018).

4.5.4 Modernizing Truck Regulation

The relationship between the gross vehicle weight of a truck and its fuel consumption is not one to one. An increase in a truck's size and payload leads to a smaller proportionate increase in fuel consumption. In other words, larger trucks with heavier payloads haul each unit of freight with less fuel than smaller trucks, all else remaining equal.

Several countries have restrictions on truck size and weight. These have been put in place mainly to limit wear and tear on roadways and bridges and to address safety concerns.

There is some momentum for revising constraints in favor of frameworks that permit so-called high-capacity vehicles without compromising infrastructure durability or safety. Political authorities could consider to introduce the so-called performance-based standards to replace current limits on vehicle weights and dimensions with design criteria that ensure that vehicles operate as desired on roadways. In particular, these standards do not impose restrictions based solely on the physical attributes of a vehicle, such as its weight and dimension; by contrast they introduce specific performance criteria in common operational settings.

Over the last decade, Australia has tested performance-based standards with the purpose to abandon the traditional truck regulation which hinges upon physical attributes like weight. The Australian standards mandate that vehicles are able to meet specific performance criteria, such as on low-speed support paths, gradeability, and rearward assistance (IEA, 2017b).

For countries adopting these kinds of standards, it is crucial to limit the risks of counterproductive, modal shift from rail to road transport. In fact, performance-based standards increase flexibility and reduce costs connected to goods shipment operated by trucks, which, as previously mentioned, largely contributes to CO_2 emissions. Given this risk, authorities should make sure that the introduction of standards designed on performance is accompanied by the adoption of measures promoting the switch from road to cleaner transport modes.

4.5.5 Promoting Alternative Fuels for Trucks

Efforts to make road freight transport cleaner include the switch to less polluting fuels. With respect to passenger cars, this switch has barely been applied due to the parallel, significant surge of electric vehicles. By contrast, truck electrification proceeds at a slower pace as batteries required by this kind of vehicles are still costly and bulky. As electric trucks hardly gain market share, fossil fuels greener than petrol or diesel, such as natural gas, are considered as a promising solution to embark on a decarbonization (or at least carbon reducing) process.

Measures to promote a fuel switch involve taxation and investments on infrastructure. For example, some countries levy lower rates of duty on natural gas than on gasoline or diesel. In the UK, the fuel duty on compressed and liquid natural gas and biomethane is 50% below that on diesel (The Oxford Institute for Energy Studies, 2019, p. 10). Also, governments encourage investments on refueling stations through subsidies to operators. In Europe a number of companies have built facilities part-funded by the EU's "Connecting Europe Facility for Transport" program (European Commission, Innovation and Networks Executive Agency). Uniper is planning eight NGV refueling stations in Germany, three in Belgium and three in France; while, in 2018, Liquind 24/7 GmbH announced (LNG World News, 2018) it would be investing €16 million in ten LNG stations following a grant of over €3 million.

The growing economy of Shenzhen, China, has raised the prominence of environmental issues. In an attempt to promote alternative fuels for transport, Shenzhen has the largest number and the biggest terminals in China. Once all terminals have become fully operational, the supply capacity of LNG will exceed 11 million tons/year accounting for around 55% of the Guangdong province (Hu, Huang, Cai, & Chen, 2017).

4.5.6 Supporting Digitalization of Railways

The rapid development involving digital technologies can pave the way for digitalizing key services linked to rail transport, thereby improving the efficiency of freight transport operated by rail. In fact, digitalization enables operators to modernize freight information exchange, enhance online traffic monitoring and detect imminent defects or breakdowns to prevent accidents. Also, digital innovations offer new solutions such as smart infrastructure and automated train control systems, favoring train automation. Higher level of automation means that trains are increasingly driven by computers which optimize their speed and control the braking systems. As a consequence, line capacity will improve.

The digitalization of railways entails considerable benefits in terms of safety, punctuality, and energy consumption. In other words, it makes freight transport more efficient and environmentally sustainable at the same time.

In order to boost the process of rail digitalization, the European Commission included the digital agenda to its 2020 strategy (European Commission, Europe, 2020 Strategy). Furthermore, in 2015, it established the digital transport and logistic forum (European Commission, Digitalization of Transport and Logistics), a collaborative platform, where Member States, public entities, and organizations exchange knowledge and coordinate policies and technical recommendations for the European

Commission, in the fields of transport and logistics digitalization across all modes of transport. This group put forward proposals such as "encouraging the use and acceptance of e-freight information by state authorities and business operators in all transport modes, including rail, and to propose interoperable IT solutions to exchange this information" (European Parliament, 2019b, p. 4).

On the funding front, the EU legislator bodies are evaluating the 2021–2027 Digital Europe program (European Commission, 2018) in the multiannual financial framework. This program is meant to strengthen the EU's high-performance computing and data processing capacities, reinforce core artificial intelligence capacities and expand the best use of digital technologies in key sectors such as transport.

Digitalization of railways should be a policy priority given the substantial and far-reaching benefits. Alongside environmental benefits there are also significant efficiency benefits to be exploited. Much funding for digitalization will still be allocated toward R&D; however, certain strategies, such as those relating to the exchange of information, can already be implemented in practice.

4.6   Transport Policies and Governance Levels

		City level	Country level	International level
Passenger Transport	Carpooling	X		
	Car sharing	X	X	
	Public transport promotion	X	X	
	Supporting connected and automated mobility		X	X
	Congestion charging	X		
	Parking management	X		
	Cycling and walking zones promotion	X		
	Multimodality	X		
	License plates limitation	X	X	
	Emission standards		X	X
	Bans on diesel and petrol vehicle commercialization		X	

(continued)

(continued)

		City level	Country level	International level
	Public investments in clean vehicles R&D		X	X
	Public investments in clean vehicles infrastructure	X	X	
	Clean vehicles production quotas		X	
	Public procurement for clean vehicles		X	X
	Subsidies	X	X	
	Non-fiscal incentives	X	X	
	Bans on diesel and petrol vehicle circulation	X		
Freight Transport	Subsidies to freight transport on railways		X	X
	Support to high-speed train		X	X
	IMO regulation			X
	Promotion of alternative fuels	X	X	
	Truck regulation		X	
	Digitalization of railways			X

4.7   Taxation

Taxation plays an essential role when considering policies to decarbonize the transportation sector. Providing a comprehensive, global overview of taxes linked to the automotive industry is complex for two reasons. First, a large variety of heterogeneous taxes are levied on the motor vehicle segment. Direct taxation includes taxes on vehicle acquisition (VAT, sales tax, registration tax), ownership (annual circulation tax, road tax) and motoring (fuel tax). This is complemented by indirect taxation, of which parking fees are an example.

Secondly, tax systems vary considerably from country to country. In fact, some national governments prefer to levy all the types of taxes on vehicle acquisition while adopting only few on ownership. By contrast, some others put greater emphasis on taxes on vehicles ownership. Certain others have significantly increased fuel taxes and, at the same time abolished some of those on vehicle ownership. Furthermore, governments can adopt the same tax, but charge different rates or complement it with other fiscal measures. For example, Germany applies VAT at 19% on the sales of new vehicles while Greece at 24%; Italy, instead, applies a 22% VAT but has introduced a bonus/malus scheme based on CO_2 emissions. Besides the levels of CO_2 emissions, other criteria contribute to differentiating rates at which taxes on motor vehicles are applied: engine's attributes (such as size, power or age), cylinder capacity, fuel type, and vehicle's attribute (weight, age, or number of axles). In sum, the tax framework is strictly dependent on national politics.

In this section, fuel taxation is first explored through cross-country comparison on the magnitude of excise taxes. Cross-country comparisons of vehicle acquisition and ownership taxes are harder to compare cross-country but insights into the EU, USA, and Chinese taxation systems are offered.

4.7.1 Fuel Taxation

The OECD has carried out some cross-country analysis with regards to fuel taxation. Figure 4.1 shows the average effective road energy tax on both diesel and gasoline. There is an apparent and significant heterogeneity between countries. The data presented are the average effective fuel taxes for both diesel and gasoline for road use by country. Significant differences in the pricing of diesel and gasoline are highlighted.

4.7.2 Acquisition and Ownership Taxation

EU Taxation

The European Automobile Manufacturers' Association (ACEA) published its annual Tax Guide which presents an overview of the specific taxation on motor vehicles in Europe. The Guide (ACEA Tax Guide, 2019) provides a summary of the taxes on acquisition and ownership of motor vehicles in all EU Member States.

Taxes on acquisition within Europe comprise of a VAT paid on initial purchase as well as a registration tax. The lowest VAT rate is 18% in Malta, while the highest is 27% in Hungary. Croatia, Denmark, and Sweden also have a notably high VAT rate at 25%.

Registration taxes across Europe vary significantly in terms of how they are computed. Typical variables used are CO_2 emissions, cylinder capacity, vehicle purchase price, and other emissions standards. For example, CO_2 emissions are used to compute payments in Austria, Belgium, Croatia, France, and Portugal among others. Cylinder capacity is taken into account for Belgium, Cyprus, Hungary, Poland, and

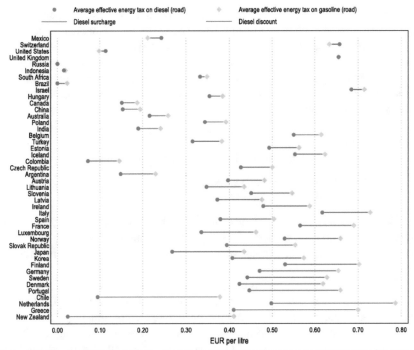

Fig. 4.1 Global overview of fuel excise taxes (OECD). *Source* OECD (2019)

Note: Average tax rates are calculated based on the tax rates applicable on 1 July 2018 and energy use data for 2016 that was adapted from IEA (2018[1]), *World Energy Statistics and Balances*. New Zealand is a special case because diesel vehicles pay distance-based road-user charges, which are not included in TEU because they affect different behavioural margins than energy taxes.

Portugal. The purchase price or value is considered by Croatia, Denmark, Finland, Greece, Ireland, and Malta among others.

The UK imposes a flat 55 GBP registration fee for all types while Sweden is the only EU country without a registration tax. In France, the registration tax varies by region and includes an additional CO_2 component. In Germany registration fees are 26.3 EUR. In Italy, the registration fee is based on vehicle type and horsepower, as well as CO_2 emissions. In Spain, registration fee is based on CO_2 emissions.

Taxes on ownership differ for passenger cars and commercial vehicles. For passenger cars, engine power and CO_2 emissions are two commonly used metrics. For commercial vehicles, gross vehicle weight and number of axles are two commonly used metrics (i.e., the size of the vehicle is the important factor in determining tax).

Engine power determines part or all of the passenger cars ownership tax in Austria, Bulgaria, Croatia, Czech Republic, and Hungary. CO_2 emissions are considered in Cyprus, Finland, Greece, Ireland, Luxembourg, and Malta among others. In France, the fiscal power and CO_2 emissions are considered. In Italy, power, emissions and fuel type are all important. Estonia, Lithuania, Poland, and Slovenia have no ownership taxes on passenger cars.

Gross vehicle weight is used in calculating commercial vehicles ownership tax for Czech Republic, Estonia, Finland, Italy, Lithuania, the Netherlands, and Poland.

For Germany, as well as weight and pollution, noise category is also taken into account. In Luxembourg, the suspension type is a component in price setting alongside weight and axles. The UK considers dead weight and environmental characteristics. Spain considers only payload in determining tax. Slovenia has no ownership tax on commercial vehicles.

China Taxation

Taxation in China can be grouped into two primary categories:

- Taxes to be paid by the vehicle manufacturer
- Taxes to be paid by the vehicle owner

Zongwei, Yue, Han, and Fuquan (2017) provide an overview of the general structure of China's taxation system as it is applied to automobiles, detailing which taxes are due at which stage of a motor vehicle's life cycle and whom they must be paid too.

Unlike within the EU, VAT is charged throughout the production life cycle of the motor vehicle rather than simply at point of sale. The first phase of VAT is therefore paid by manufacturers and indirectly passed through to consumers.

There are four phases of taxation in the Chinese system: producing, purchasing, retaining, and using. A variety of taxes are paid by automakers in the producing phase. These include a circulation tax which varies between 1 and 40% depending on engine. VAT is paid, split between central and local governments. Surtaxes are levied on urban maintenance and construction and education. An estate property tax and city and town land use tax must also be paid.

The rest of the taxes for stages purchasing, retaining, and using are all paid directly by consumers. Consumers pay a 10% tax of the vehicle price when purchasing to the central government. Local government's then charge VVT which is a local tax varying by engine displacement and applied for the retaining of an automobile.

The final stage of taxation comes from using an automobile. Circulation taxes are paid on both gasoline and diesel. These are approximately $0.25/L for gasoline and $0.2/L for diesel. VAT, UMCS, and ES are all paid to the central government. UMCS and ES are surtaxes of VAT on refined oil.

USA Taxation

In the USA, most vehicle taxes are imposed at the state rather than federal level. One exception is the "gas guzzler tax." This tax was established through the Energy Tax Act of 1978 and applies to the sale of new vehicles whose fuel economy fails to meet certain statutory requirements. Its purpose is to discourage the purchase of fuel-inefficient vehicles.

The tax code has also recently been utilized to promote the purchase of electric vehicles and plug-in hybrid vehicles through providing tax credits to purchasers.

Such vehicles purchased in or after 2010 may be eligible for a federal income tax credit. The minimum credit amount is $2500 rising to $7500 depending on traction battery capacity and the gross vehicle weight rating.

The credit is applied until the manufacturer has produced 200,000 eligible vehicles counted from January 1, 2010. The IRS announces when a manufacturer exceeds this production figure and will announce the phase out schedule (for the tax credit). As of October 2019, tax credits have been phased out for both Tesla and General Motors.

Most US states impose a sales tax on vehicle purchases. This is a tax on consumption, displayed as a percentage of the sale price. All US states impose vehicle registration fees. These fees are imposed on all vehicles regardless of age and are generally collected annually (ACEA Tax Guide, 2019).

4.8 Risks Linked to Digitalization

Digitalization plays a key role in several of the policies examined. On the one hand, modern digital technologies enhance public transport efficiency, make carpooling and car sharing services more accessible to users, support multimodality and pave the way for smart cities and grids which, in turn, represent the essential prerequisite for automated and interconnected vehicles. On the other hand, increasing levels of digitalization, which we notice inter alia in the spread of the so called Internet of things[2] (IoT), entail potential risks in terms of security.

As digitalization growingly pervades our everyday life, potential targets of cyber-attacks evolve and become more numerous. Cyber-attacks have predominantly targeted IT systems operating in computers and smartphones. Yet, as our everyday objects are increasingly connected to digital networks, they become vulnerable to intrusions from hackers.

This threat involves all sectors, including transport. As one possibility, one could imagine the consequences deriving from a cyber-attack targeting a software which regulates automated cars and trucks. Hackers could stop vehicles' circulation or, in worse cases, trigger accidents.

Aside from security risks, digitalization presents privacy concerns. Privacy issues arise because the functioning of digital technology requires large amounts of personal data. For example, every time we buy bus tickets online, book a car sharing trip or use bike sharing services we provide operators with personal data. We share our starting point and our destination. Also, in the case we carry out the same trip regularly, we share important information regarding our daily routine (where we live, places that we and our relatives frequent, time at which we move, routes we follow). Such data could be used for commercial data mining. That is, customers might be exposed to intrusive targeted marketing techniques or even criminal activities.

---

[2]With this expression we refer to the situation "where everyday objects are connected to networks to provide a range of services or applications in areas such as cars, home automation and smart grids." IEA (2017a).

It is highly likely that automated and interconnected vehicles will require even more detailed personal data. Therefore, the privacy-linked concerns are set to rise in the future.

Authorities have stepped up efforts to address challenges posed by digitalization. In order to face the cyber security issues, the EU adopted the Network Information Security Directive (Directive (EU) 2016/1148) in 2016, requiring Member States to develop a national strategy. This directive established a co-operation group to facilitate strategic cooperation between the Member States regarding the security of network and information systems. As regards privacy, a new regulation, GDPR (Regulation (EU) 2016/679) was drafted to protect personal data. The GDPR states that "consent should be given by a clear affirmative act establishing a freely given, specific, informed and unambiguous indication of the data subject's agreement to the processing of personal data." Moreover, it sets rules regarding data breach notifications.

4.9   Distributional Effects of Policies

This final section sheds light on the potential distributional effects deriving from policies to foster the decarbonization of the transportation sector. Specifically, it discusses the social risks linked to fiscal measures pursued to discourage the use of conventional internal combustion engine vehicles as well as potential solutions. Such measures include the introduction of a carbon (or fuel) tax and the reduction of fossil fuel subsidies.

Although environmentally friendly, these measures, as any other fiscal measure, entail wealth distribution which, in turn, may result in social protests. Carbon pricing and similar fiscal initiatives particularly involve major hazards, as they have serious repercussions on a large part of the population, including low-income classes. The case of France will help identify key aspects of this issue.

The French carbon tax was introduced in 2014 at a rate of €7/tCO$_{2e}$.[3] Within four years, it rose sixfold, thus reaching the rate of €44.6/tCO$_{2e}$. Spurred by the necessity to comply with the Paris agreement requirements, as well as the desire to step up efforts toward an effective ecological transition, the French government proposed a further reform in 2018. The new carbon tax was set on a rising price trajectory toward €86.2/tCO$_2$ in 2022 and accompanied by an increase in fuel prices.

This proposal soon triggered serious social turmoil. Strongly opposing these measures, the "yellow vests" movement spontaneously emerged and spread throughout the country. Demonstrations were not only motivated by direct and indirect increase in fuel prices, but also (and mainly) by a perceived discrimination. In the yellow vests' opinion, higher fuel costs predominantly affect workers from the countryside. In fact, these people massively consume fuel because, on average, they drive

---

[3]Unless otherwise stated, all data related to France carbon tax refer to World Bank Group, State and Trends of Carbon Pricing (2019c).

longer distances than workers in cities. Beside the geographical (countryside-city) discrimination, this proposal gave rise to lower-income people's concerns, as they cannot necessarily afford modern, cleaner vehicles. Their concern is that they will have to carry the tax burden alone, since wealthier people gradually shift toward new generation, lower CO_2 emitting cars.

As a consequence of large scale protests, the French government decided to modify (and partially withdraw) its plan: the tax rate in 2019 will remain at the 2018 rate of €44.6/tCO$_2$. As the yellow vests' riot is still continuing at the time of writing despite nation-wide consultations, a new attempt to increase the carbon tax in the near-term is highly unlikely.

Aside from the French case, there are examples where governments successfully introduced plans to support an ecological transition without neglecting dangerous social implications. Carbon pricing or reforms to reduce fossil fuel subsidies were implemented together with cash transfers (or other similar measures) to most vulnerable social classes with the purpose to mitigate inequalities. Such combined plans were able to bring benefits on two different fronts.

In 2008, British Columbia, Canada unilaterally introduced a broad-based carbon tax applying to carbon emitted from the combustion of all greenhouse gases, with few exemptions granted. The tax began at 10 Canadian Dollars (CAD)/tCO$_2$ and rose to 30 CAD/tCO$_2$ by 2012. The tax was implemented in a revenue neutral manner with measures designed to ensure equitable effects. Revenue from the tax was used to reduce the bottom two personal income tax rates and the low-income climate refundable tax credit was introduced returning annually up to $100 per adult and $30 per child. Furthermore, all residents received a one-time $100 dividend aimed at contributing toward lifestyle restructuring away from carbon emitting sources (British Columbia, 2008). The combination of these measures ensured the political survival of the tax through attempts to alleviate some of the worst distributional consequences.

Recently, Jordan and Iran have embarked on a process to lower fossil fuel subsidies while, in parallel, developing benefit programs to counter poverty effects. In 2012, the government of Jordan "used various social safety net measures to protect vulnerable groups, including cash transfers to low-income households, a targeted food subsidy program, and increasing the public sector wages for lower income households" (World Bank Group, 2019c, p. 78). Similarly, "Iran implemented electronic cash transfers that accounted for 50% of projected savings from fossil fuel subsidies reforms" (Ibid.).

In sum, the experiences in France, British Columbia, Jordan, and Iran show two essential aspects related to the topic of this chapter. First, fiscal policies to foster decarbonisation are likely to bring about distributional effects. Given the nature of the policies, these distributional effects tend to disadvantage lower-income classes. Second, measures increasing carbon or fossil fuel prices do not necessarily imply growing inequality. If complementary fiscal programs are implemented, two key aims can be achieved at the same time, thereby benefiting both environment and population.

4.10   Conclusions and Key Take-Aways

This chapter has looked to offer a broad overview of the policies available to ensure effective decarbonization of transport sectors. There are a wide range of policies that can be implemented. Individual countries will have particular and specific advantages/disadvantages with regards to the implementation of each individual measure. There is therefore no perfect policy combination for ensuring effective decarbonization of the transport sector but the aforementioned policies are the key areas in which countries have tried, are trying, and must continue to try to invoke positive climatic change.

One common theme emerging from the chapter is of the wide range of positive externalities associated with decarbonization policies. These policies primarily have an impact on reducing carbon emissions however so many are found to impact society in a multitude of other positive ways. For instance, measures aimed at a reduction of road transport demand lead to cleaner cities in terms of air pollution, while developments in automated technologies lead to wide-ranging efficiency benefits. Meanwhile, measures such as the digitalization of railways and promotion of multimodality not only reduce carbon demand but can ensure a smoother and more enjoyable transit experience for passengers.

In terms of precise policy implementation, there is no one-size-fits-all strategy. In order to implement successful decarbonization policy mixes, this chapter has shown that policymakers must pursue a range of subsidies, fines, regulatory measures and standards, bans, R&D investments, management strategies as well as cross-border cooperation.

One area stands out in which governments should look to adopt an effective policy mix. Certain policies are aimed at directly reducing the occurrence of polluting activity, e.g., emission regulations and city- or country-level bans. These policies, by design, will be effective. However, governments should be careful at the same time to pursue policies that will provide alternatives for citizens to switch too once the status quo is banned. R&D and subsidies for promising but not yet developed technologies should be followed in order to develop zero emissions technologies. Meanwhile, the provision of infrastructure, such as ZEVs charging networks, is pivotal in order to encourage private investment to follow. There is a careful balance to be struck between forcing citizens to change behavior and at the same time providing suitable alternatives.

At the same time, measures to promote and inform the general public of the benefits of action are fundamentally important. Government policies should be transparent, for example, information campaigns can explain how any revenue accruing is to be used such that decarbonization policies do not become confused with governments simply trying to raise revenues.

Beyond promoting awareness of the benefits of climate policy, it is of paramount importance to recognize and confront the potential realities of significant socioeconomic consequences. Whether it is measures to subsidize new technologies, accessible mainly by higher-income classes (e.g., EVs) or taxes on fuel consumption (where the incidence is often largest on lower-income classes) such policy measures are often regressive. In order to combat regressive effects and ensure broad population participation and enjoyment of the energy transition within the transport sector, governments must consider this issue of the upmost importance. Distribution of revenues in a progressive manner is a good start, but furthermore governments must look to assess the regressive impacts of individual policies and factor these into any decision-making process.

References

5G Carmen. Available at gcarmen.eu.

5G Croco, https://5gcroco.eu/.

5G PPP. Available at https://5g-ppp.eu/5g-mobix/.

Abalate, D., & Bel, G. (2012). High-speed rail: Lessons for policy-makers from experiences abroad. *Public Administration Review, 72*.

ACEA Tax Guide. (2019). Available at https://www.acea.be/uploads/news_documents/ACEA_Tax_Guide_2019.pdf

BBC. (2020). *Petrol and diesel car sales ban brought forward to 2035*. Available at https://www.bbc.com/news/science-environment-51366123.

Bloomberg. (2019a). *A ban on dirty shipping fuel is coming. so why are prices surging?* Available at https://www.bloomberg.com/news/articles/2019-04-23/with-demand-poised-to-collapse-the-price-of-ship-fuel-rockets.

Bloomberg. (2019b). *In Beijing, you have to win a license lottery to buy a new car*. Available at https://www.bloomberg.com/news/articles/2019-02-27/in-beijing-you-have-to-win-a-license-lottery-to-buy-a-new-car.

British Columbia. (2008). Budget and Fiscal Plan 2008/09–2010/11.

Brussels Capital Region, Low Emissions Zone. Available at https://www.lez.brussels/.

Center on Global Energy Policy. (2019). Electric vehicle charging in China and the United States.

City of Copenhagen. (2011). *The City of Copenhagen's bicycle strategy 2011–2025*. Available at https://www.eltis.org/sites/default/files/case-studies/documents/copenhagens_cycling_strategy.pdf.

City of Copenhagen. (2017). *Copenhagen City of cyclists*. Available at http://www.cycling-embassy.dk/wp-content/uploads/2017/07/Velo-city_handout.pdf.

CIVITAS Initiative, Public Transport Promotion Campaign. Available at https://civitas.eu/measure/public-transport-promotion-campaign.

CIVITAS Initiative, Roma. Available at https://civitas.eu/city/roma.

CIVITAS Initiative, Turku. Available at https://civitas.eu/eccentric/turku.

CIVITAS Initiative. *Upgrading the car-pooling system with an events feature*. Available at https://civitas.eu/it/measure/upgrading-car-pooling-system-events-feature.

CURACAO. *What are the key features and examples of scheme design?* http://www.isis-it.net/curacao/?content=keyscheme.

Department of Environmental Affairs, Republic of South Africa. (2014). Freight shift from road to rail.

Department of Transport, Republic of South Africa. (2017). National rail policy: Draft white paper.

Deutsche Gesellschaft fur Internationale Zusammenarbeit (GIZ). (2018). Defining the future of mobility: Intelligent and connected vehicles (ICVs) in China and Germany.

Directive (EU) 2016/1148 of the European Parliament and of the Council of 6 July 2016 concerning measures for a high common level of security of network and information systems across the Union. Available at https://eur-lex.europa.eu/legal-content/EN/TXT/PDF/?uri=CELEX:32016L1148&from=EN.

European Commission. 5G Public Private Partnership, the Next Generation of Broadband infrastructure. Available at https://ec.europa.eu/digital-single-market/en/news/5g-public-private-partnership-next-generation-broadband-infrastructure.

European Commission. *Digitalisation of transport and logistics and the digital transport and logistics forum*. Available at https://ec.europa.eu/transport/themes/logistics-and-multimodal-transport/digitalisation-transport-and-logistics-and-digital-transport-and_en.

European Commission, Europe 2020 strategy. Available at https://ec.europa.eu/digital-single-market/en/europe-2020-strategy.

European Commission, Innovation and Networks Executive Agency. Available at https://ec.europa.eu/inea/en/connecting-europe-facility/cef-transport.

European Commission, Mobility and Transport. Available at https://ec.europa.eu/transport/modes/rail/packages/2013_en.

European Commission, Project closer. Available at https://trimis.ec.europa.eu/project/connecting-long-and-short-distance-networks-efficient-transport#tab-outline.

European Commission. *Reducing CO₂ emissions from passenger cars*. Available at https://ec.europa.eu/clima/policies/transport/vehicles/cars_en#tab-0-0.

European Commission. *State aid: Commission approves €350 million per year in public funding to promote shift of freight transport from road to rail in Germany*. Available at https://europa.eu/rapid/press-release_IP-18-6747_en.htm.

European Commission, Trans-European Transport Network (TEN-T). Available at https://ec.europa.eu/transport/themes/infrastructure/ten-t_en.

European Commission. (2018). *Proposal for a Regulation of the European Parliament and of the Council establishing the Digital Europe programme for the period 2021–2027*. Available at https://eur-lex.europa.eu/legal-content/EN/TXT/?uri=COM%3A2018%3A434%3AFIN.

European Commission—Directorate General for Mobility and Transport. (2017). Delivering TEN-T. Available at http://www.connectingeu.eu/documents/Delivering_TEN_T.pdf.

European Parliament. (2019a). *Review of the clean vehicles directive*.

European Parliament. (2019b). *Digitalisation in railway transport: A lever to improve rail competitiveness*.

European Union Agency for Railway. (2018). *The 4th Railway Package—What does it mean for me?* Available at https://www.era.europa.eu/sites/default/files/library/docs/leaflets/4th_railway_package_what_does_it_mean_for_me_en.pdf.

Gao, G., Jiang, C., & Larson, P. (2017). Express freight transportation by high-speed rail: The case of China. In *Proceedings of the 52nd Annual Conference Canadian Transportation Research Forum*.

Government of National Capital Territory of Delhi Transport Department. (2019). *Draft Delhi maintenance and management of parking places rules 2019*. Available at http://transport.delhi.gov.in/content/notification-draft-delhi-maintenance-and-management-parking-places-rules-2019-sought.

Governor Edmund G. Brown Jr., ZEV Action Plan. (2018). Available at http://business.ca.gov/Portals/0/ZEV/2018-ZEV-Action-Plan-Priorities-Update.pdf.

Greentech Media. (2018). *California regulators open a new chapter in utility EV charging policy*. Available at https://www.greentechmedia.com/articles/read/california-regulators-open-a-new-chapter-in-utility-electric-vehicle-chargi#gs.1bdt4m.

Guardian. (2015). *Congestion charge has led to dramatic fall in accidents in London*. Available at https://www.theguardian.com/uk-news/2015/mar/07/congestion-charge-accident-fall.

Guardian. (2019). *Amsterdam to ban petrol and diesel cars and motorbikes by 2030*. Available at https://www.theguardian.com/world/2019/may/03/amsterdam-ban-petrol-diesel-cars-bikes-2030.

Hu, M., Huang, W., Cai, J., & Chen, J. (2017). The evaluation on liquefied natural gas truck promotion in Shenzhen freight. *Advances in Mechanical Engineering, 9*(6).

Institute for Transportation and Development Policy. (2018). *China tackles climate change with electric buses*. Available at https://www.itdp.org/2018/09/11/electric-buses-china/.

Integrated Transport Centre, DARB, Abu Dhabi. Available at https://www.darb.ae/Carpooling/Home/index.

International Association of Ports and Harbors, IAPH LNG bunkering audit tool already used to license operations at Port of Rotterdam: LNG bunkering operations set to grow exponentially in next five years. Available at http://www.iaphworldports.org/news/5424.

International Council on Clean Transportation. (2018). *China's new energy vehicle mandate policy (Final Rule).*

International Energy Agency (IEA). (2017a). *Digitalization & Energy.*

International Energy Agency (IEA). (2017b). *The Future of Trucks Implications for energy and the environment.*

International Maritime Organization, Sulphur 2020—Cutting sulphur oxide emissions. Available at http://www.imo.org/en/MediaCentre/HotTopics/Pages/Sulphur-2020.aspx.

Islam, D. Z., Ricci, S., & Nelldal, B. (2016). How to make modal shift from road to rail possible in the European transport market, as aspired to in the EU Transport White Paper 2011.

Joint Research Centre. (2019). Best environmental management practice for the public administration sector. Available at https://susproc.jrc.ec.europa.eu/activities/emas/documents/PublicAdminBEMP.pdf.

Keall, M. D., et al. (2018). Reductions in carbon dioxide emissions from an intervention to promote cycling and walking: A case study from New Zealand. *Transportation Research Part D, 65.*

LNG World News. (2018). *EU funds German LNG fueling project.* Available at https://www.lngworldnews.com/eu-funds-german-lng-fueling-project/.

Lubrizol. (2019). *China's ever-tightening fuel consumption regulations.* Available at https://www.lubrizoladditives360.com/chinas-ever-tightening-fuel-consumption-regulations/.

Lutsey, N., Searle, S., Chambliss, S., & Bandivadekar, A. (2015). Assessment of leading electric vehicle promotion activities in United States cities.

Marine and Port Authority of Singapore. (2019). *IMO 2020 sulphur limit—A guide for bunkering industry.* Available at https://www.mpa.gov.sg/web/portal/home/port-of-singapore/services/bunkering/imo-2020-fuel-oil-sulphur-limit.

Marine and Port Authority of Singapore, LNG Bunkering (Pilot Programme). Available at https://www.mpa.gov.sg/web/portal/home/port-of-singapore/services/bunkering/lng-bunkering-pilot-programme.

Maritime Executive. (2019). *2019 will be the year of acceleration for LNG as Marine Fuel.* Available at https://www.maritime-executive.com/editorials/2019-will-be-the-year-of-acceleration-for-lng-as-marine-fuel.

Menon, G., & Guttikunda, S. (2010). *Electronic road pricing: Experience & lessons from Singapore.*

Metro-magazine. (2015). *The 'Uber Effect': Will new ride services reinvent transit.* Available at https://www.metro-magazine.com/mobility/article/410225/the-uber-effect-will-new-ride-services-reinvent-transit.

New York Department of Transportation. (2018). Final Report: Truck Platooning Policy Barriers Study.

Nottingham City Council. *Workplace Parking Levy.* Available at http://www.nottinghamcity.gov.uk/wpl.

OECD. (2019). *Taxing Energy Use 2019: Using taxes for climate action.*

Polinomia srl. (2016). *Andamento della mobilità ciclistica a Milano.* Available at http://www.polinomia.it/assets/files/bollettino_resoconto_2016.pdf.

Port of Rotterdam. *Study reveals LNG reduces shipping GHG emissions by up to 21%.* Available at https://www.portofrotterdam.com/en/news-and-press-releases/study-reveals-lng-reduces-shipping-ghg-emissions-by-up-to-21.

RailFreight.com. (2018a). *High-speed freight train Italy hits the track on 7 November.* Available at https://www.railfreight.com/railfreight/2018/11/02/high-speed-freight-train-italy-hits-the-track-on-7-november/?gdpr=accept&gdpr=accept.

RailFreight.com. (2018b). *ISC launches a second high-speed rail freight service in Italy.* Available at https://www.railfreight.com/business/2018/10/23/isc-launches-a-second-high-speed-rail-freight-service-in-italy/.

Regulation (EU) 2016/679 of the European Parliament and of the Council of 27 April 2016 on the protection of natural persons with regard to the processing of personal data and on the free movement of such data, and repealing Directive 95/46/EC, available at https://eur-lex.europa.eu/eli/reg/2016/679/oj.

Report on National R&D Programmes on the Fully Electric Vehicle. Available at https://trimis.ec.europa.eu/sites/default/files/project/documents/20140203_102303_58862_D_2.4_Report_National_R_and_D_Programs_Final.pdf.

Reuters. (2018). Denmark embraces electric car revolution with petrol and diesel ban plan. Available at https://www.reuters.com/article/us-denmark-autos/denmark-embraces-electric-car-revolution-with-petrol-and-diesel-ban-plan-idUSKCN1MC121.

Safety4Sea. (2019). *Port of Rotterdam reports reduction in fuel oil, increase in LNG.* Available at https://safety4sea.com/port-of-rotterdam-reports-reduction-in-fuel-oil-increase-in-lng/.

Slowik, P., & Lutsey, N. (2017). Expanding The Electric Vehicle Market In U.S. Cities.

The Guardian. (2016). *Four of world's biggest cities to ban diesel cars from their centres.* Available at https://www.theguardian.com/environment/2016/dec/02/four-of-worlds-biggest-cities-to-ban-diesel-cars-from-their-centres.

The Guardian. (2018). *Shenzhen's silent revolution: world's first fully electric bus fleet quietens Chinese megacity.* Available at https://www.theguardian.com/cities/2018/dec/12/silence-shenzhen-world-first-electric-bus-fleet.

The Hindu. (2019). *Dear Delhi, is it finally time to ditch you car?* Available at https://www.thehindu.com/news/cities/Delhi/is-it-finally-time-to-ditch-your-car/article29551188.ece.

The National. (2019). *Would you give a lift to a stranger? Abu Dhabi rolls out new carpooling system.* Available at https://www.thenational.ae/uae/transport/would-you-give-a-lift-to-a-stranger-abu-dhabi-rolls-out-new-carpooling-system-1.922597.

The Oxford Institute for Energy Studies. (2019). *A review of prospects for natural gas as a fuel in road transport.*

United States Mid-Century Strategy FOR DEEP DECARBONIZATION. (2016). Available at https://unfccc.int/files/focus/long-term_strategies/application/pdf/us_mid_century_strategy.pdf.

Wiredbugs, 10 Countries Banning Fossil Fuel Vehicles Before 2050, available at https://wiredbugs.com/countries-banning-fossil-fuel-vehicles/.

World Bank Group. (2019a). China's high-speed rail development.

World Bank Group. (2019b). The rail freight challenge for emerging economics.

World Bank Group. (2019c). State and trends of carbon pricing.

Zongwei, L., Yue, W., Han, H., & Fuquan, Z. (2017). Overview of China's Automotive Tax Scheme: Current Situation, Potential Problems and Future Direction. *Journal of Southeast Asian Research.*

Printed in the United States
By Bookmasters